Praise for Elizabeth Royte's
Garbage Land
On the Secret Trail of Trash

"Fascinating. . . . Royte is a dogged reporter and a vivid writer, which means that her catalog of crimes against nature hits the senses hard. She has a keen eye and a sensitive nose."
— William Grimes, *New York Times*

"An eye-opening, nose-holding, and ultimately disturbing odyssey through the waste of a nation. . . . Royte is an entertaining writer . . . and turns what could be dry policy or shrill tub-thumping into an engaging, often darkly funny tale."
— Thomas Maresca, *Milwaukee Journal Sentinel*

"Captivating. . . . As impressive as Royte's doggedness and investigative skill is the care she takes with language. In a book where facts and figures are so plentiful and ominous, felicitous phrasing can work like the proverbial spoonful of sugar. . . . Royte never lets us forget that she is an ordinary citizen just like the rest of us, coping with the demands of marriage and motherhood while chasing her story. . . . Her frequent returns to the many scenes in her tiny kitchen help ground her expansive narrative in a way that keeps it fresh — in a manner of speaking — and accessible."
— Jabari Asim, *Washington Post Book World*

"An eye-opening ride through the culture of waste disposal."
— Bruce Barcott, *Outside*

"Engaging. . . . A fascinating story. . . . A personable glimpse into what happens to the stuff we bung into the bin."
— Elizabeth Grossman, *Portland Oregonian*

"A riveting travelog punctuated by a scathing indictment of American consumption." — Daniel Terdiman, *Wired*

"Utterly fascinating. . . . The land of garbage is evidently a war zone, and this book is an astonishing, if pungent, dispatch from the front lines."

— Anthony Brandt, *National Geographic Adventure*

"This is the most comprehensive (and funny) look by far at a subject of perennial interest: where does it all go? Any city dweller will read it with fascination and mild horror; if you've ever survived a New York City garbage strike, every smell will come unbidden back to your flaring nostrils."

— Bill McKibben, author of *Wandering Home*

"For a dose of sympathy and a fascinating education, pick up *Garbage Land*. . . . The author recounts her experiences with a dose of humor that makes it easier to swallow the sobering statistics about our trash. . . . *Garbage Land* is not just for tree huggers. Anyone who rolls a trash cart out to the curb once a week will appreciate Royte's inquisitiveness. . . . Amid all the data, Royte uses her sharp analytical ability to step back and contemplate the nature of our trash in thought-provoking, philosophical terms."

— Heather Landy, *Star-Telegram*

"Surrounded by sobering statistics, Royte is a modern-day, modernist muckraker, exhibiting more irony, realism, and resignation than righteous indignation."

— Glenn C. Altschuler, *Boston Globe*

"Alive with observation."

— Nicholas Fonseca, *Entertainment Weekly*

"This book stinks! It also entertains, illuminates, frightens, and inspires. Part rollicking road trip, part reconnaissance from the scary front lines of the ecological sciences, *Garbage Land* takes us deep down into a ninth circle of our own making to reveal the maggoty truths of our throwaway culture."

— Hampton Sides, author of *Ghost Soldiers* and *Americana*

"Royte takes a simple, smelly idea and blows it up large. She's conveyed some of the wonkier debates over waste management while bringing out the human side of a story most humans would prefer to ignore." — John Dicker, *Now*

"Pungent. . . . Blunt and unadorned. There is anger and often a touch of humor."

— Diane Roberts, *Atlanta Journal-Constitution*

"Eye-opening." — Moira Bailey, *People*

"The best book ever about trash. . . . Royte knows how to orchestrate telling statistics and vivid descriptions to illuminate every dirty corner of the business. . . . Highly readable and exhaustively documented." — Jim Motavalli, *Grist*

"An excellent excursion into our ephemera and rejectamenta, both of which say more about us than we ever seem to understand."

— Robert Sullivan, author of *Rats* and *The Meadowlands*

"Royte approaches her subject with a sense of irony and a gift for narrative. . . . Indispensable."

— Howard Hall, *Time Out New York*

"Recently, a new class of muckrakers has investigated the hidden infrastructures that provide us with the thing we want—from fast food to fossil fuels. With *Garbage Land,* Elizabeth Royte reverses the formula, shrewdly exposing the mechanisms by which 'our unwanted stuff keeps disappearing.' That is, she rakes actual muck. . . . Royte scrutinizes with a sharp eye and a pinched nose."

— John Mooallem, *Mother Jones*

"Royte's exploration of the economic, territorial, and ecological perspectives of garbage disposal adds up to a fascinating trail of trash. Recommended for all who throw things away."

— Irwin Weintraub, *Library Journal*

Garbage Land

Also by Elizabeth Royte

The Tapir's Morning Bath:
Solving the Mysteries of the Tropical Rain Forest

Garbage Land

ON THE SECRET TRAIL OF TRASH

Elizabeth Royte

BACK BAY BOOKS
Little, Brown and Company
NEW YORK BOSTON LONDON

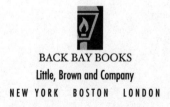

Back Bay Books / Little, Brown and Company
Hachette Book Group USA
1271 Avenue of the Americas, New York, NY 10020
Visit our Web site at www.HachetteBookGroupUSA.com

Originally published in hardcover by Little, Brown and Company, July 2005
First Back Bay paperback edition, August 2006

The author is grateful for permission to include lyrics from "It Really Isn't Garbage Till You Throw It Away" by Danny Einbender. Words and music by Danny Einbender. Published by Mindbender Music / BMI.

Library of Congress Cataloging-in-Publication Data

Royte, Elizabeth.
Garbage land : on the secret trail of trash / Elizabeth Royte.—1st ed.
p. cm.
ISBN 0-316-73826-3 (hc) / 0-316-15461-x; 978-0-316-15461-x (pb)
1. Refuse and refuse disposal—New York (State)—New York. I. Title.
HD4484.N7R68 2005
363.72'85'097471—dc22 2004024732

10 9 8 7 6 5 4 3 2 1

Q-FF

Book design by Marie Mundaca
Printed in the United States of America

For Peter and Lucy

Contents

Introduction

Quantifying in the Kitchen 3

Part One: To the Dump

One Dark Angels of Detritus 27
Two Amphibious Assault 50
Three Stalking the Active Face 63
Four The Spectacle of Waste 85

Part Two: Avoiding the Dump

Five Behold This Compost 105
Six Forward into the Flexo Nip 127
Seven Hammer of the Gods 142
Eight Mercury Rising 158
Nine Satan's Resin 176

Part Three: Flushing It Away

Ten Downstream 197
Eleven In the Realm of Taboo 210

Part Four: Piling On

Twelve It's Coming on Christmas 235

Thirteen	The Dream of Zero Waste	251
Fourteen	The Ecological Citizen	277
	Acknowledgments	297
	A Note on Sources	299
	Index	301
	Reading Group Guide	315

Garbage Land

Introduction

Quantifying in the Kitchen

On a sunny spring afternoon long before I ever decided to travel around with my garbage, I slid off the dead end of Second Street, in the Boerum Hill neighborhood of Brooklyn, and down a seven-foot embankment oozing green and brown liquid. I braced my foot on the end of a rotting nineteenth-century beam and prayed that it would hold. It did, and soon I was seated in a slime-encrusted canoe in the Gowanus Canal, my sneakers awash in bilgewater. My life vest and jeans now bore distinctive parallel skid marks. A sportsman in a Gowanus Dredgers cap released the bowline and casually informed me that those row houses—he pointed up Second Street—were discharging raw sewage into the canal. "That would explain the smell," I said.

It was Earth Day 2002, and I'd come out not to collect floating garbage—the siren call for two dozen local Sierra Club members—but to get a little exercise. I'd never paddled around the city, and I wanted a new perspective on my neighborhood. I also wanted a backyard view of what the media were touting as up-

and-coming real estate. "Gowanus," after morphing into the tonier-sounding "Boerum Hill" in the sixties, was returning as a sales category.

I left the proffered dip net and trash bucket on the embankment and turned the canoe deeper into Brooklyn. It was low tide, and the smell was, even for someone expecting the worst, fairly bad—a combination of outhouse, mudflat, and mold. The water was a diarrheal brown and topped by a slick of psychedelically swirled oil. I J-stroked past a fuel oil depot, a sewage outfall pipe, and the tin-can-cluttered encampment of a hobo. I glided over submerged shopping carts coated thick with algae and watched as other paddlers plucked spent condoms—or Coney Island whitefish, as they're locally known—from the water's surface. It occurred to me, as I turned to work my way out toward Gowanus Bay, that I was paddling through a microcosm of the city's multifarious effluent. In one small, horribly polluted, godforsaken stretch of water drifted household trash, raw sewage, toxic waste, containers that ought to have been recycled, and rapidly putrescing organic debris. With a start, I realized it was all the stuff I got rid of almost daily.

Scanning the canal and its collapsing bulkheads, I wondered if I was complicit in this specific mess. I lived uphill, in Park Slope, and understood that garbage has a tendency to roll down, to settle on the margins. Before this day I'd wondered only idly how my detritus disappeared. You can't live in New York or any big city and not be aware that vast tonnages of waste are generated daily. If you're unlucky enough to be around during a garbage strike or an extended snow emergency, those tonnages assume a visceral reality. But most of the time that reality is virtual, because somehow our unwanted stuff keeps disappearing. It moves away from us in pieces—truck by truck, barge by barge—in a process that is as constant as it is invisible.

Now, as I paddled slowly through the Gowanus feculence, my curiosity grew. I understood that my regular trash went to some kind of landfill, but what about my recyclable tuna fish cans and my plastic shampoo bottles? These containers were tipped into the same truck, but surely the combined waste streams were at some point teased apart. Where, and by whom? And then what? My waste was no longer within my sight or smell, but surely it fell within others'. What impact did my rejectamenta have on other living things? Once I started to think about these questions, I couldn't let them go.

I felt drawn to the Gowanus for atavistic reasons (who doesn't like the shore?), but I was also interested in the canal as a backyard conduit and as a junkyard, of sorts. Over the years, the Gowanus had developed a reputation as a dumping ground for the mob; a character in Jonathan Lethem's *Motherless Brooklyn* refers to the canal as "the only body of water in the world that is 90 percent guns." Some of New York's garbage infrastructure was overt, some was covert, and the Gowanus seemed to fall somewhere in between. The canal was one hundred feet wide and one point eight zigzagging miles long, not counting the three spurs, called basins, that led to the loading docks of warehouses and factories along the avenues. There were enormous gravel barges tied to the canal's edge and sunken barges sitting on its bottom. Among the facilities that actually made use of the water were an asphalt plant (which used to incorporate the city's recycled glass into "glassphalt"); a marine transfer station, where the borough's residential garbage had, until a year ago, been tipped into barges bound for Staten Island's Fresh Kills landfill; a couple of cement factories; and two fuel oil companies. A guy named Orion lived on a houseboat near Carroll Street, and Lenny "the Chicken Man" Thomas worked atop the Union Street Bridge, raising the drawbridges when tallish boats requested it and developing recipes for street-cart barbecue when they didn't.

When the Dutch first arrived in Brooklyn, in the early part of the seventeenth century, the Gowanus was a tidal creek that ran through the salt marsh valley between Park Slope, where I lived, and Carroll Gardens. (The word *Gowanus* comes from the Iroquois chief Gowanes.) Native Americans lived well on fin- and shellfish they collected in the briny waters. The Dutch farmed the local oysters and exported them by the barrelful to Europe—"oysters as big as dinner plates," Owen Foote, a cofounder of the Gowanus Dredgers Canoe Club, said to me (everyone who talks or writes about New York City oysters uses dinnerware as a measuring stick).

Gradually, dams and landfills altered the salt marsh's ecology. In 1849, the New York State Legislature authorized construction of a straightened and walled canal. (Of course, the Gowanus wasn't really a canal, since it didn't connect anything, but creeks didn't qualify for state construction money.) South Brooklyn was rapidly becoming industrialized, and the canal, completed in the late 1860s, was soon lined with stone yards, flour mills, chemical plants, cement works, and factories that turned out paint, ink, and soap. By then, Brooklyn was America's third-largest city. Barges hauling brownstone and bluestone, lumber and brick—the stuff that built my apartment house and its environs in the 1880s—jostled for position in the harbor, waiting to enter the canal.

Almost from the beginning, the Gowanus was a filthy place: with limited tidal exchange to open water, the discharge of raw sewage, combined with unregulated industrial waste, stagnated. Local residents, appalled by the stench, launched a campaign for improvement. In 1911, the city completed construction of the Gowanus Flushing Tunnel, designed to suck 200 million gallons of water a day from the East River, flush it for more than a mile underneath Brooklyn, and then discharge it into the back end of the canal. The neighborhood celebrated the pump's opening on a June afternoon with Miss Gowanus gliding up the canal on a barge, strewing flower petals in her wake. In the years following World War I, the Gowanus moved six million tons of material a year: it was the nation's busiest commercial canal.

But it didn't last. After World War II, the Gowanus Expressway was opened, and trucks gradually siphoned work from barges. The canal became a stinky anachronism: the pump broke in 1961, and there was no money for repairs. In 1989, South Brooklyn got a wastewater treatment plant, but it did little to solve the Gowanus's odor and pollution problems. Through twelve wastewater overflow pipes, storms still deposited raw sewage and toxic rainwater into the canal. Finally, a decade after the treatment plant opened, the pump was fixed. The city dredged two thousand tons of contaminated muck from the canal's bottom, and the Gowanus Dredgers and the Urban Divers, another community group focused on canal restoration, weighed anchor.

The busy summer paddling season passed, and Ludger K. Balan, the Urban Divers' self-styled environmental program director, offered me a private ecology tour of the canal. We arranged to meet, in two days, at the end of Second Street. I showed up early, then watched the appointed hour, from my perch atop a concrete slab, come and go. The tide slowly dropped and the sun beat down. The water was so clear I could make out a chaise longue settled peacefully on the canal's bottom. Condom wrappers and Colt 45 tallboys littered the boatyard. Was there no place off the beaten track that was free of this stuff? Only slightly annoyed, I basked in the autumnal warmth and made note of my surroundings. "I Peed on U," someone had scrawled on the Dredgers' equipment locker. "Hey Fear You Blue-haired FAG."

I checked my watch for the third time. In the weeks and months ahead, I'd learn that time was loosely constructed in Balan's world, that directions were vague, phone numbers often garbled, e-mails so badly written it was difficult to tell if events were planned or were already history. The Gowanus Dredgers ran a tighter ship, but it was Ludger K. Balan who had offered me an ecology tour, and so it was Ludger K. Balan's club that I paid twenty bucks to join.

At long last, Balan pulled up in a van decorated inside and out with plastic fish and mermaids. A short, muscular black man, he wore neoprene booties, rubber bracelets, a brown sweater under red rubber overalls, and a woolen headband wrapped around a teapot-sized bun of dreadlocks. Balan was half Haitian and half Arab, he said. He had grown up in France "and four other countries." I noticed that his British accent came and went as he warmed or cooled to his subject.

We drove up Bond Street to a weedy, potholed lot. This was the Divers' actual boatyard, a corner of a plot owned by a local named Danny, who seemed to be at the center of several smallish operations involving heavy equipment, the film industry, and auto-body repair. Balan and I wended past refrigerators, pipes, iron beams, and movie trailers, then carefully tiptoed across the top of the bulkhead toward two aluminum skiffs painted orange. We manhandled one onto a floating dock and loaded it with wooden oars and a five-gallon bucket filled with water-sampling instruments. Balan, a self-appointed waterway steward, claimed to collect data twice a week, in good weather, and hand it over to the Army Corps of Engineers, which was considering dredging the canal once again. The Urban Divers organized scuba trips around the city and spent significant energy on public education, which included luring in potential "citizen monitors," like me.

Settled in the boat now, Balan rowed us past the Bayside Fuel Oil company, where giant oil tanks were buried beneath grassy berms, a hedge against explosion. Bayside was the only company that brought boats this far back into the canal these days, but it couldn't take oil deliveries at low tide or during extreme temperatures, which sometimes interfered with the smooth operation of the drawbridges.

An underwater filmmaker, Balan had come of age in a hundred-foot visibility zone. "Diving in the Caribbean was almost overstimulating," he said. "In the Hudson, you can see just three feet. It's quiet and meditative." He occasionally dove in the Gowanus, but it was an ordeal. He wore complete protective gear, including a rubber dry suit with a face mask. It took a long time to put everything on. "None of your body parts can touch the water," he said. "When you get under, you try not to disrupt the sediment, for visibility and for health." Afterward, everything had to be meticulously washed. "We have spotted poo-poo in the canal," he added in the tone of a TV anchor. In July of 2000, the Brooklyn Center for the Urban Environment had optimistically planted five thousand caged oysters in the canal as an indicator of water quality: today, only eighty stunted survivors remained.

At the Third Street Bridge, we hauled up a fish trap and counted four fillies, grayish minnows about two inches long. Once, Balan found 115 in a single trap. On other days, the Urban Divers had hauled up silversides, toadfish, tomcod, sea robins, flounder minnows, and pipefish. I wrote "4 fillies" in Balan's log and started handing him his tools. The water temperature today was 15 degrees Celsius. The pH was 7.5—a little acidic. Salinity was 23. To estimate water clarity, we lowered a Secchi disk—a white plastic circle about the size of a Colonial-era Gowanus oyster—and got a reading of four feet. Anything over five feet was considered pretty good, with declines in transparency typically due to high concentrations of suspended solids: sediment, plankton, and the aforementioned poo-poo.

Next, Balan dropped a stylus filled with electrolytes over the gunwale. "We could be in dead water," he said gravely. "A normal dissolved oxygen level is five point eight parts per million. We've got point two." I suspected faulty equipment, but Balan suspected the pump. "Dissolved oxygen levels drop in about a day when it's broken," he said.

The occasional failure of the pump is, according to some, intentional. If the canal becomes too clean, certain businesses may no longer be welcomed here. Today, the canal is a sacrifice area, a series of brownfields zoned for industry, and not a few manufacturers want to keep it that way. The pump is at the heart of the matter. When it's broken, floating debris and chemical spills aren't flushed; when it's operating, everything looks and smells better. Many canal activists credit the pump with bringing wildlife back to the canal. First came the oxygen, then plankton, then fin- and shellfish (oysters, mussels, and crabs), and then waterfowl. In 2002, Balan's group documented thirty-eight species of birds around the canal, and a couple of Jet Skiers in it, too. But like a federally listed endangered species to a strip mall developer, the idea of a cleaner, greener Gowanus is anathema to some.

A few months ago, Balan and his wife, Mitsue, had collected eighteen large bags of trash from a grassy patch between the canal and the Pathmark supermarket, by the Hamilton Avenue bridge.

They had asked the store to take the full bags, but they'd refused. So did the Brooklyn South 6 sanitation garage, even though it was just one avenue away and the guys were over here constantly (the Dunkin' Donuts adjacent to the supermarket was open twenty-four seven). Eventually, the Balans themselves hauled away the sacks. The tiny lawn they had cleaned was now a carpet of vibrant green shadowed by a birch in full autumnal splendor. It would have been a nice place to sit and look at the water, but for the racket of traffic overhead.

A Columbia University agronomist once told me that coffee-drinking habits in the New York metro region had the potential to affect the hillsides of faraway coffee-growing nations. We ran through a lot of beans in the city, she explained: almost 204 million pounds a year, based on a conservative average of 1.7 cups per person per day. If all those beans were grown in, say, El Salvador, they'd dominate the country's harvest. Of course, New Yorkers bought beans from many different countries, but the professor had made her point. The choices we make have repercussions far and wide. Buying shade-grown coffee that conserves forests for other species and supporting fair labor practices could have a salutary effect on people and places we'll never see.

William Rees and Mathis Wackernagel, regional planners in Canada, developed the ecological footprint concept as a way to measure the sustainability of our lifestyles. Basically, a footprint totals the flows of material and energy required to support any economy or subset of an economy (coffee drinking, for example), then converts those flows into the total land and water surface area that it takes to both provide those resources and assimilate their waste products. For residents of densely populated cities, that surface area extends well beyond our borders, into the hinterlands. We don't grow much, and our water and energy come from afar. Measuring our coffee footprint, or any other footprint, isn't necessarily about good and bad; it is about making informed choices.

But lattes were just the beginning. Mindful of our consumptive lifestyles, I imagined the city had a garbage footprint bigger than any in the world. We were eight million people, we consumed and threw out a lot, and we had very little nearby space in which to dump our discards. For nearly fifty years, New York City relocated its trash—a peak of thirteen thousand tons a day from houses and apartments, plus an additional thirteen thousand tons a day from commercial and institutional buildings—to the Fresh Kills landfill on Staten Island, our least-populated borough. In 1986, Fresh Kills became the largest dump in the world. It rose two hundred feet above its surrounding wetlands and formed the highest geographical point along fifteen hundred miles of eastern seaboard.

Fresh Kills closed in March of 2001, and for the first time in its history, the city had no place within its boundaries to bury or burn all the stuff its residents no longer wanted. Now the city exports almost all its solid waste to outlying states. Our footprint, which has always been big, has suddenly become a lot bigger. And New York isn't the only city spreading its garbage toes.

Since 1960, the nation's municipal waste stream has nearly tripled, reaching a reported peak of 369 million tons in 2002. That's more stuff, per capita, than any other nation in the world, and 2.5 times the per capita rate of Oslo, Norway. The increase is due partly to increased population but mostly to the habits of average residents, who now throw out, says the EPA, 4.5 pounds of garbage per person per day—1.8 more pounds than forty-five years ago. According to the Congressional Research Service, the biggest producers are California, followed by New York, Florida, Texas, and Michigan. *BioCycle* magazine and the Earth Engineering Center of Columbia University reported in their "State of Garbage in America" report for 2003 that every American generated 1.31 tons of garbage a year. Nearly 30 percent of the aggregate mess was recycled or composted, according to the EPA; 13 percent was incinerated; and the overwhelming majority, 57 percent, was buried in a hole in the ground.

After paddling the Gowanus, I became increasingly curious to learn what sort of impact my own 1.31 annual tons had as it meandered through the landscape. To do that, I had to go on a garbage tour, of sorts. But before I started my far-flung travels with trash, I decided to acquaint myself, like Thoreau in Concord, with the extremely local, and take a close look inside my own kitchen waste bin. Like fossils, ancient kitchen middens, and Clovis points hewn by early man—evidence scrutinized by scientists peering into our past—the stuff we reject today reveals a great deal about human beings and how they live. What would my garbage say about me? What exactly was I throwing out, and how much of it was there?

My voyage of self-discovery, like so many voyages, began with an acquisition. I unwrapped my Polder 2 in 1 Gourmet Add N' Weigh Digital Scale & Kitchen Timer and settled the white plastic disc on my counter. Battery-operated and sleek, it could handle a maximum of seven pounds. "Surely I won't generate more than seven pounds of trash a day," I told myself naively. An empty wine bottle, I'd learn that very night, weighs about one pound.

I couldn't wait to begin digging through my garbage. After dinner I collected my tools on the kitchen floor: the scale, a notebook, a pen, an empty plastic bag. I sat down and tightly tied the full trash bag—a plastic grocery sack—to keep it from toppling off the scale. After weighing, I untied the sack and started removing items one by one, writing down their names and placing them in the new bag. This sounds straightforward, but it wasn't. My pen got sticky; coffee grounds spilled onto the floor. My daughter, Lucy, who was three at the time, was instantly at my side offering help. Halfway through, I washed my hands and put on a pair of rubber gloves, which made writing difficult. My data for the first day looked like this:

> *October 3. Foil packaging from Fig Newmans, empty box of sandwich bags, waxed paper bag from muffin shop, 2 plastic bags from vegetables, plastic bread bag, coffee grounds, receipt from grocery store, grapefruit and watermelon rinds, misc. food scraps from dinner, 1 slice stale bread, 1 banana peel, 5 basil stems, 1 half-gallon plastic milk bottle, 2 half-gallon juice cartons, 1*

beer bottle, 1 jelly jar, 1 wine bottle, 1 half-liter plastic bottle of chocolate milk, 1 peanut butter jar, miscellaneous "fines."

Total weight: 7 pounds, 9 ounces.

I was a little embarrassed about the contents of my trash, especially the chocolate milk container. It had been a treat for Lucy. I didn't usually buy individual servings of anything: they were expensive and their packaging created more trash. William Rathje, founder of the University of Arizona's Garbage Project, which was established by archaeologists to study both human discard habits and the inner dynamics of landfills, insists that refuse reflects truth. Garbage sorting reveals that "what we do and what we say we do are two different things." We underestimate how much booze we drink; we overestimate our leafy greens. I resolved to be more careful about chocolate milk containers, though I reckoned I'd have a hard time explaining it to Lucy.

I returned to my diary entry. *Fines,* a word used by Rathje in his garbage sorts, included floor sweepings, dust, strands of hair, coffee grounds—all the tiny stuff that settled to the bottom of the bin. I noted the cardboard box from the sandwich bags. That had been a mistake: I'd been too lazy to bring it out to the paper-recycling pile on my landing. I didn't feel so bad about the beer bottle: I had weighed it, but it wasn't going to the landfill because New York was a bottle bill state. I'd put it on the sidewalk for a homeless guy named Willy, who'd redeem the container at the local beverage center for the nickel deposit.

I couldn't have begun quantifying my garbage at a more confusing time in New York's recent history. If I'd started my project four months earlier, I would have been recycling four pounds, one ounce of my total weight (or 51 percent) and been sending just three pounds, eight ounces to the landfill for two days in the life of my small family. But years ago, our Republican mayor Rudolph Giuliani had promised the overwhelmingly Republican residents of Staten Island that he would close Fresh Kills. Now, instead of paying about $40 a ton to dump waste within the city limits, we paid

$105 a ton to export it. Facing a tight budget, our current mayor, Michael Bloomberg, had recently suspended the recycling of glass and plastic, claiming that it cost too much to collect and process. City environmentalists were outraged, and so was I. My project had barely begun and already it was complicated by politics.

I didn't want to let this blip in the history of New York skew my data. Bloomberg had promised that plastic recycling would return in one year and glass in two. (He was persuaded by environmentalists to keep his hands off the metal and paper streams, which continued to bring revenue to the city.) I considered putting my project on hold, if only so I'd have a chance of beating the national average. The decision to include or exclude became morally freighted. Ultimately, I decided to weigh my glass and plastic separately, just so I'd have the data, then total my garbage both with and without these materials.

When I was done combing through my trash, I put a new bag in the empty kitchen can and brought the full sack down to the mother bin in my brownstone's front yard. Then I washed my hands and reviewed the first night's lessons. I noted that food waste, the wet stuff, really messed up my garbage. That wine bottles were heavy. That Peter, my husband, had thrown away a hunk of moldy cheese that he could easily have trimmed (if he didn't have a phobia about mold). That I had left small bits of paper in my trash. That I could probably do a much better job of shrinking my garbage footprint.

There was something else I noticed, too. The plastic sack with which I'd just lined my trash can was no longer empty. I'd turned my back for five minutes, and already the waste was accumulating. Was there no relief from it? Did the flow ever stop? I wondered if sanitation workers ever felt a sense of futility. They cleaned one street after another after another, until the district was officially clean. But no sooner were the bins tipped than they immediately began to fill. Emptiness—cleanliness—was a condition so brief as to be nearly undetectable. "You can't think about that," one of my sanitation workers (or san men, as both men and women called themselves) told me. "You'll drive yourself crazy."

In two days' time, the kitchen bag was full. Again, Lucy sat on the floor next to me, wearing a rubber glove that was twelve sizes too large. Coffee grounds speckled her thigh as she sorted plastic from glass and held up objects for identification.

"Having fun?" I asked.

"It's a little smelly," she answered. "Daddy, what's this?" She held aloft something soft and red.

"That's a chicken liver," Daddy answered from his position at the stove.

Turning toward me, Lucy asked, "Why do we have so much trash?"

I gave her the proximate answer. "Because we keep throwing things out."

"Why do we throw things out?" She handed me a plastic milk cap and answered herself. "Because they get yucky." Did she mean yucky before they hit the can or after? I sang a little song to her: "It really isn't garbage till you mix it all together. / It really isn't garbage till you throw it away. / Just separate the paper, plastic, compost, glass and metal, / And then you get to use it all another day."

"Mommy," she said, "will you sing 'Stewball'?"

After a week of sitting by my side, Lucy lost interest in combing through garbage. She was lucky to live here, in the grand ol' USA. In many developing nations, entire families pick through municipal dumps together in search of materials—fabric, metals, paper, glass—that can be exchanged for cash. The work is hierarchical: the highest-ranking families have rights to the most valuable stuff, usually metals. Rooting through my garbage, I wondered briefly which items I'd keep if I had to live off this waste. Then I realized I wouldn't have bought most of this stuff in the first place, or thrown most of it out, had I been in that position.

I was now dumping the trash on Lucy's blue plastic toboggan before sorting it, and that kept the floor cleaner. The sled was a mess, though, and it took a few tries till I learned how to rinse it in the kitchen sink without dumping a quart of water on the floor. I considered buying a composter for all the food waste, or at least the coffee grounds, which coated everything in the kitchen bag. But I didn't have a garden in which to use the finished product, and cultivating rotting food outside my brownstone would surely alienate my neighbors. Or so I thought at the time.

As the Garbage Project discovered, "Garbage expands so as to fill the receptacles available for its containment." (Project researchers called this Parkinson's Law of Garbage, after the original law formulated by C. Northcote Parkinson, a British civil servant based in Singapore: "Work expands so as to fill the time available for its completion.") My house had one trash can in the kitchen, a tiny one in the bathroom, and two more in bedrooms. By making it easy to toss things away, was I was abetting garbage mindlessness?

It's hard to imagine, but 125 years ago the kitchen trash can didn't exist. Until municipal collections were organized, in the late 1880s, the stove was the principal means of disposal. But the oven door wasn't opening and closing all day long, like a kitchen trash can. Food scraps went to farm animals. Individually packaged consumer goods were rare and expensive. Tin cans were saved for storage or scoops, jars for preserving food. Old clothes were repaired, made over into new clothes, or used for quilting, mattress stuffing, rugs, or rags. Plastic was unknown. As late as 1882, reports Susan Strasser in *Waste and Want: A Social History of Trash*, a manual on teaching children household economy had to *define* a wastebasket for readers: "It is for collecting all the torn and useless pieces of paper, and should be emptied every day, care being taken that nothing of value is thus thrown away."

But what was valued? In the days of household economy manuals, almost all castoffs and scraps could be used as barter. Today, my aluminum cans had cash value to a scrap metal dealer in New Jersey, but my wine bottle, which the city no longer recycled, was dead weight in the garbage truck. Those fourteen ounces were still a commodity, though: the more weight the city buried in landfills, the more money landfill owners pocketed.

There are other types of value assigned to trash. Artists see beauty in certain forms of litter; parents of preschoolers imbue their offspring's every mixed-media collage with sentimental value. For composters, organic waste is a treasure trove of nitrogen. To some, litter is a tool of anarchy.

Most Americans keep multiple wastebaskets in their homes, but I decided to quantify only my kitchen trash. In the interest of full disclosure: my bedroom trash was almost entirely paper. I tried to write on both sides, then recycled the larger pieces. The bathroom trash was used tissues, stubs of soap, dental floss, and, once a month, evidence of menstruation. Now and then I'd empty the little bathroom basket into the kitchen bag, but it added at most a few ounces. I didn't mind picking through my own used tissues, but I had little interest in picking through others', even those of people I love. And here was another universal garbage truth: other people's waste is always worse than your own.

> *October 10. Two plastic wrappers from magazines, plastic from cheese, plastic from a bill of lading, 1 plastic box from fresh pasta, 1 Ziploc of slimy parsley, 1 plastic box from Fig Newmans, 1 foil-lined paper bag from mint Milanos, Lucy art (tempera on paperboard), Lucy art (collage of wax paper, tinfoil, and Saran wrap), 1 half-gallon juice carton, 1-gallon plastic milk bottle, 1 paper towel (used to clean up previous garbage sort), 6 paper napkins, 1 plastic tape dispenser, 2 stained cloth napkins, 2 stained place mats, apples, coffee grounds, onions, green-pepper trimmings, pea pods, lots of grapes, spoiled cherry tomatoes, fines.*
>
> *Total weight: 4 pounds, 2 ounces.*

Every week or so I had cause to throw out some sort of textile, and each time I jotted it down, history jabbed me in the ribs. As someone minding her footprint, I ought to have saved my stained napkins for rags. But rags weren't scarce in my house: I had enough to wash every window in the neighborhood. I realized that using old clothes or napkins as disposable dust rags merely postponed their trip to the landfill. So why do it? Because using a rag meant I wasn't using a paper towel, which spared a fraction of a tree from being milled and a fraction of a river from some toxic papermaking discharge.

A century and a half ago, I might have saved my stained shirtwaists for the local peddler, who sold textiles to paper companies. Peddlers also relieved households of ashes, old metal, bones, and rubber, delivering them to soap manufacturers, tinsmiths, button- and boot-makers. The peddlers, in turn, supplied housekeepers with manufactured goods. This two-way trade—the earliest form of household recycling—allowed housewives to acquire goods without cash, and it was essential to the development of certain industries in the mid-nineteenth century.

Returning raw materials to manufacturers to be refashioned into other goods looked like Yankee thrift. But it was also a form of nascent consumerism, Strasser notes, a way to acquire products not grown or otherwise created at home. My situation, a hundred and fifty years on, was just the opposite. Like most people, I tended to do right by the environment—whether avoiding disposables or scrupulously turning off lights—mostly when it saved me money.

Picking through garbage was smelly and messy and time-consuming, but it was revelatory in a way. I hadn't realized my diet was so boring. Anyone picking through my castoffs would presume my family survived on peanut butter, jelly, bread, orange juice, milk, and wine. And, largely, we did. It occurred to me late one night, as I sat peacefully on the floor surrounded by the remains of the day, that I knew something about where all this stuff had come from (particularly if it was food; the nation's heightened health consciousness inspired a lot of ink on the provenance of foodstuffs) but almost nothing about where it went after it left my house. Much has been made, in certain circles, of humanity's connection to the natural world. Enlightened consumers, we don't want to eat endangered fish or buy rare hardwoods. We care about animal rights and clean water. But it wasn't fair, I reasoned, to feel connected to the rest of the world only on the front end, to the waving fields of grain and the sparkling mountain streams. We needed to cop to a downstream connection as well. Our lifestyles took a toll on the planet, and that toll was growing ever worse.

> *October 24. One* Jane Goodall's Wild Chimpanzees *video, 1 plastic shopping bag, 1 plastic bread bag, 1 plastic veggie bag, 1 cardboard egg carton (not in paper recycling because there's a broken egg in it), 5 paper towels (from cleaning up broken egg), 2 one-pint ice cream containers and tops, Saran wrap, 1 bakery bag with leftover bialys, 1 butter paper, 4 plastic lids from coffee cups (would a careful observer surmise, from the lack of coffee grounds, that the household ran out of coffee and for two days purchased lattes, at twice the cost of a pound of coffee beans?), 1 foil packet of soy sauce, half a peanut butter sandwich, carrot peels, onion skins, lemon rind, 1 chicken carcass, soggy chicken bedding in Styrofoam tray, couscous, orange rind, bread, fines.*
>
> *Total weight: 10 pounds, 6 ounces, of which 7 pounds, 8 ounces is recyclable (7 wine bottles, 1 half-gallon juice carton, 1 one-liter plastic bottle from olive oil, 2 jars).*

The only time I really dreaded quantifying garbage was after dinner parties. I waited until the last guest was gone, then wearily hunkered down on the kitchen floor to extract the items that less footprint-minded friends had tossed into the trash: the brown bag from a wine bottle (to the recycling pile), a rubber band from a bouquet (to the odds-and-ends drawer), a beer bottle (to Willy).

Parties made my kitchen garbage wet, heavy, and smelly. I blamed the meat. A hundred years ago, I'd have handed over my leftover chicken carcass—probably with far less flesh on it—to the local "swill children," who supplied rag- and bone pickers with material that they in turn sold for buttons and knife handles. The fat rendered from bone marrow would have gone to factories for lighting and lubricating; gelatin was used in making glue and in processing food and photographs. Bones also made excellent fertilizer, a commodity that became increasingly important to farmers as untilled land became scarce after the turn of the nineteenth century. If I wanted to recycle my chicken here at home, I could have made candles from the grease now coating the trash bag, or soap, by combining it with lye that I made by dripping water through wood ashes. (For one Martha Stewart-ish moment I considered this. Like many brownstones in my neighborhood, mine had a working fireplace. But I'd need several pounds of ashes to get started, and the heating season was young.)

Looking at the postprandial mess arrayed before me, I assumed that I was generating far more waste today than I would have fifty or a hundred years ago. For one thing, there were no triple-wrapped Fig Newmans or mint Milanos back then. But in fact, if calculated by weight alone, I was doing pretty well. According to Daniel C. Walsh, a professor at Columbia University's Department of Earth and Environmental Engineering who examined a century's worth of city garbage records, per capita rejection in New York peaked in 1940 at 2,068 pounds a year, or 5.66 pounds per day. It dropped to a century-low 712 pounds a year in the midseventies (the economy was in poor shape) and by 1999 rose to 928 pounds. The rate, he reported in *Environmental Science & Technology,* has been fairly steady since 1980. (Walsh attributed this nearly flat line to reductions in the weight of bottles and cans, and the advent of deposit bills. We may actually have been throwing out more, he implied, but the more weighed less.)

The big difference between then and now is our fuel sources. Approximately 34 percent of the 3.5 billion tons of refuse generated over the twentieth century was coal and wood ash. Looking just at the century's first four decades, the ash fraction was even higher: a full 60 percent. By 1950, the use of coal for heating had declined to the point where paper replaced it as the largest proportion of the city's residential waste stream. I laughed when I read Walsh's hopeful prediction that paper would ultimately be replaced by more economic digital technologies: I'd read his paper online, then printed it out.

According to Walsh, the mass fraction of food in the (ash-free) waste stream dropped from 65 percent in the early 1900s to 13 percent in 1989, thanks to improved refrigeration, the increased use of chemical preservatives that lengthened shelf life and reduced spoilage, and the increased availability of frozen food, which resulted in the sale of fewer untrimmed vegetables. While food waste went down, however, packaging waste went up. Americans had become more prosperous, and thanks to the evolution of technology behind consumer goods, there was far more stuff available for them to buy.

The advent of different types of plastic, between the thirties and the forties, radically altered how Americans kept house. Polystyrene made refrigerators more affordable, for example, and Plexiglas reduced the cost of manufacturing headlights, lenses, windows, clocks, and jewelry. Manufacturers began hyping disposable products—sanitary napkins, paper towels, plastic cups—as scientific, modern, and hygienic. Tapping into class prejudices, ad campaigns suggested that the old ways, linked to poverty and recent immigration, were dirty. (A Kotex ad in 1927 claimed that "80% or more better-class women have discarded ordinary ways for Kotex." "Ordinary ways" were reusable cloths.) The new disposables were touted as time- and labor-savers that would boost women into the leisure class. To resist the siren call of the new, writes Strasser in *Waste and Want,* was to risk being branded backward and fearful.

Before New Yorkers burned or buried their waste, they pitched garbage out their windows and onto the city streets, where it was consumed by scavenging pigs and dogs. It was the same in any large American city. Still, there was always more refuse than animals, swill children, and ragpickers could handle. By the 1800s, the filth in lower Manhattan had accumulated to a depth of two to three feet in the wintertime, when household waste and horse manure combined with snow. My brownstone in Park Slope, like others built in the late 1800s, has a stoop leading to the second floor, which let residents clamber above the mess (though it still seeped into the ground floor during storms and when snow melted). For much of the nineteenth century, trash removal was a private, not municipal, service, which made garbage an issue of social class. I don't know who lived in my building a hundred and twenty-odd years ago, but it's likely they paid someone to take their ashes and food scraps away, to be dumped with other wastes into the Atlantic Ocean.

Periodically, but usually spurred by outbreaks of disease, city officials made concerted efforts to clean the streets. It wasn't a simple matter. Even when Manhattan's population was less than a million, in the mid-nineteenth century, city horses dumped 500,000 pounds of manure a day on its streets, in addition to 45,000 gallons of urine. These were hardworking beasts, and their average life span was just two and a half years. In 1880, according to historians, 15,000 dead horses had to be cleared from city streets. A single carter couldn't lift a horse, so the carcasses often lay around until scavengers and the elements reduced their mass. At this point they were unceremoniously tipped into the river, along with household refuse, or sold to "reduction plants" on Barren Island, out in Jamaica Bay, where they were steamed and compressed to produce grease, fertilizer, glue, and other unguinous by-products.

In 1895, a reform mayor ousted Tammany Hall, Manhattan's popular Democratic political machine, and appointed a crusading new commissioner of street cleaning, Colonel George E. Waring Jr. Working under the auspices of the Health Department, Waring put an end to sporadic cleanup efforts, instituted regular trash pickups, and required New Yorkers to separate their garbage into three curbside bins for fuel ash, dry rubbish, and "putrescible" waste (this quaint label for the wet stuff is still used by the Department of Sanitation today, though it now refers to anything that's headed for the dump).

The putrescibles were barged to the reduction plants and the ash delivered to landfills. (Brooklyn's was carted to Fishhooks McCarthy's smoldering Corona ash dump, in Queens, which became the model for F. Scott Fitzgerald's Valley of Ashes, "a fantastic farm, where ashes grow like wheat into ridges and hills and grotesque gardens." The ash dump closed in 1933; six years later, the World's Fair rose on its site.) As it had been for many years, dry garbage, after being picked clean of valuable materials like rags and paper, was used to fill waterways and wetlands, creating tens of thousands of acres of valuable waterfront real estate, including most of lower Manhattan, the Red Hook shoreline of Brooklyn, and almost the entire northern and southern fringes of both Kings and Queens Counties, upon which our airports were built.

New Yorkers in 1895 were just as balky about separating garbage as New Yorkers are today, and Colonel Waring's diversion rate (that is, the amount of stuff he kept out of landfills) was not high. In 1898, Tammany Hall recaptured the mayor's office, ended the recycling program, and resumed ocean dumping. The garbage killed oyster beds and it interfered with shipping. When waterfront-property owners complained about animal carcasses and rags on their beaches, the city once again dialed back ocean dumping (though it wasn't banned by the federal government until 1934), and a single stream of unsorted garbage flowed to eighty-nine open dumps scattered around the boroughs.

By the forties, public tolerance for the accumulating filth and vermin reached a tipping point. The city responded by closing its festering mounds and opening incinerators. At one point, twenty-two so-called burn units (in addition to the scores of small-scale "toasters" stoked by superintendents in high-rise apartment buildings) operated throughout the city, spewing noxious black smoke into the skies. The haze was so thick at times that Manhattan couldn't be seen from New Jersey.

As the small dumps were phased out and incinerators fell into disfavor, the city pioneered other methods of entombing waste. In the newfangled "sanitary" landfills, garbage was covered with a blanket of dirt at the end of each working day. The dirt muffled odors and kept vermin at bay (that is, if it was applied soon enough. In Santa Marta, Columbia, buzzards gorging on unburied trash have become too fat to fly, prompting rescue efforts by environmentalists). New York's first modern dump was Robert Moses's Fresh Kills, which opened for business in 1948. Staten Island residents weren't happy about the abrasive master builder's plan, but Moses had promised them that the landfill would close in three years and that they'd get a new highway in return for their indulgence. Moses died in 1981, twenty years before the last Fresh Kills–bound garbage barge was tugged out of New York Harbor.

The more I learned about the history of garbage in New York, the more I saw that it was a history of interim solutions, of reactions to crises political, economic, and social. Even when the federal government stepped in, change was achingly slow. Congress passed the Clean Air Act in 1970, for example, but it took until 1994 for New York City to shut the damper on its last municipal incinerators. For more than two hundred years, New York's garbage has changed hands through cronyism and favors, and landed on the backs of the disenfranchised. Only recently have NIMBY-ism and advocates for environmental justice begun to push back. Sometimes garbage is shunted elsewhere, but always at great cost.

It's the same anyplace, really. Whether you live in rural West Virginia or inner-city Chicago, you don't want other people's garbage anywhere near your backyard. Yet Americans everywhere are producing steadily more waste. Politicians devise short-term solutions, and waste managers, who own the means of disposal, seem to hold all the cards.

By the time I began traveling with my trash, Fresh Kills had been closed for two years. I knew that the city's garbage was now trucked far and wide, but I didn't know exactly where my stuff went or what happened to it once it arrived. Early one morning, I watched from my third-floor vantage point as a packer truck compacted my peanut butter jars and chicken bones with those of my many, many neighbors. What had been mine was now, unceremoniously, the city's. It was time to come downstairs, to find out what happened next.

Part One

To the Dump

Chapter One

Dark Angels of Detritus

On a cool October morning, I caught up with John Sullivan and Billy Murphy in the middle of their Park Slope garbage route. I watched them carefully, from a slight distance, but still it took me several long minutes to figure out, in the most rudimentary way, what my san men were doing. They moved quickly, in a blur of trash can dragging, lid tossing, handle cranking, and heaving. Though barrel-chested and muscle-bound, they moved with balletic precision. Sometimes Murphy and Sullivan appeared to be working independently, other times they collaborated. Save for the grunts and squeals of the truck, it all happened in relative silence. While Murphy drove to a gap between parked cars, Sullivan slid barrels up the sidewalk to the waiting truck. Sometimes Murphy jumped down to load, sometimes Sullivan did it on his own. Then they switched. The truck moved in jerks, halting with a screech of brakes. Although most sanitation workers stopped for coffee at eight, Sullivan and Murphy kept loading. Upon their return to the garage at ten-thirty, no one voiced the usual san man's query: "Did you get it up?" Sullivan and Murphy—twenty-year veterans of the Department of Sanitation, each approaching the age of fifty—had, as they always did.

I'd met the team at 6:00 a.m. roll call at the local DSNY (for Department of Sanitation, New York) garage, a low brick structure on the farthest fringe of the neighborhood. It was still dark when I locked up my bike and walked hesitantly into a large, dimly lit room filled with garbage trucks: eleven for household refuse and nine for recycling. I made my way down a cinder-block corridor lined with smoking san men and into the fluorescent-lit office. Like many a high school principal's redoubt, it had a window overlooking a hallway filled with loiterers and humming with paranoia. There were even lockers and a lunchroom down the way.

Waiting for Jerry Terlizzi, the district supervisor, to appear, I took a look around. Every stick of furniture—desks, cabinets, footlockers—appeared to have been plucked from the street and coated with the same brown paint. The walls were crammed with yellowed memoranda and notices but held not a scrap of decoration. A dark roan dog and a dull black cat padded around the building, former strays, but even their names seemed impermanent.

"The dog, the dog. Oh yeah, that's Lupo," said an officer uncertainly when I inquired. And the cat?

"Her name is Meow," answered a clerk.

"No, it's Mami," corrected another.

While I waited for Terlizzi to get off the phone and call roll, I listened to the men.

"It's gonna rain the next three days."

"Oh, man, that garbage is gonna be heavy. You're gonna lose five pounds on Friday alone."

"I hate rain. That's a drag."

"Yeah, well, you're a garbageman."

Behind me, someone said in a mincing tone, "Can I fill out a job application?" That was for my benefit, so I chuckled along with everyone else. Then two men came in from the street, jostling and punching each other's shoulders. One said, "Somebody just stole the wheels off a bike out there!" I sprang for the door, and the guys laughed.

"Just kidding, but I wouldn't leave it there. Some bum from the park is gonna steal it. Bring it in here." He said it "he-yeh." I went out to get my bicycle and when I got back briefly pretended someone had stolen the seat, prompting instant outrage, though it was actually in my backpack.

I looked at the worker cards stuck into a bar on the Plexiglas window. The rectangles of cardboard were soft with handling, inscribed mostly with Italian and Irish names, and coupled with trucks identified with an alphanumeric. As senior men, Sullivan and Murphy had exclusive day use of truck CN191 (another team would use it at night). The junior men took whatever they were handed. By now, about thirty men were standing around smoking and chatting in their dark green DSNY sweatshirts. The garage had one female sanitation worker, but she wasn't in today. When I'd meet her later, she'd invite me to use her private bathroom, which was decorated with cute animal posters.

New to this scene, I was struck by the way the men spoke to one another. They were loud and harsh, in one another's faces. They seemed quick to anger. Maybe there was too much testosterone in a small place. Or maybe just too many men who didn't like to have a boss breathing down their neck, a factor that had lured some of them to the job. Inside, the complaints never ceased. So-and-so was an idiot. The night crew never did its job right. The boss could go to hell. I'd be crushed by such contempt, but no one here seemed to mind.

Terlizzi was parked behind a small desk. He was tall and thin, with wavy silver hair, high cheekbones, and a bemused manner. "I'm missing a truck," he told a clerk, irritated. Its collection ticket, which would state how much weight the truck had tipped at the transfer station, hadn't shown up in his paper or electronic records. The clerk opened a program on the ancient computer and scrolled down. "I checked that already," Terlizzi barked. The clerk sighed, and Terlizzi stepped out to call roll.

Two to a truck, the men roared into the twilit streets, and soon the office fell quiet. After asking me to sign a waiver, Terlizzi handed me over to John Burrafato, who worked on Motorized Litter Patrol. A pugnacious man with a small black mustache and a military bearing, Burrafato cruised the district in a department sedan, making lists of bulk items—pallets discarded in an industrial area, a blown tire in the middle of the road—to be collected by truck. He noted problem areas and wrote $25 summonses to residents who didn't follow the recycling rules and $1,500 summonses (going up to $20,000) for wholesale illegal dumping. Because DSNY spent relatively little on public education, only a minority of city residents seemed to understand all the garbage rules. Being pugnacious, then, was a prerequisite for this job.

Burrafato was supposed to bring me up to my neighborhood, where Sullivan and Murphy were already at work. But he wasn't ready to do this quite yet. First, there was paperwork for him to clear, then a mechanic to insult. I sat on a brown footlocker and read the *Daily News* while he flitted in and out of the office. Terlizzi had found his missing truck, but the air was still poisoned with his ill humor. Someone on the telephone was pressuring Terlizzi to sign off on some forms. He said, "I *didn't* say when I'll do it, but if you need it right now, I'll come *back* there and *do* it!" He leaped to his feet and slammed down the phone. "Fuck *you!*" he shouted at the supplicant, who could easily have been me. Just yesterday, I had been pestering him on the phone about getting the waiver. "You make that coffee yet?" he growled at Burrafato now.

"No, I'm getting these guys straightened out." Burrafato went out, came in, went out.

"Okay, the coffee's done. You want a cup?" Burrafato was talking to me.

"No, thanks," I said, wanting only to get out of there. Burrafato went into a small room and poured himself coffee. By now I was reading the sports section. Then I read the ancient memos on the walls and studied the maps, trying to figure out my district, garage, section, and route. New York's roughly 320 square miles are broken into fifty-nine sanitation districts, where about 7,600 workers clean the streets twenty-four hours a day, six days a week.

(In comparison, Los Angeles has about 580 workers tidying up 450 square miles, but its population is less than half of New York's eight million, and its trucks host just one worker instead of two. Chicago, with a population of 2.9 million spreading over 228 square miles, relies on 3,300 sanitation workers.) Brooklyn's districts are divided into zones North and South. Here in South, there are eleven other garages besides my own, which is called the Six. The territory covered by the Six is in turn broken into five sections: my street is part of Section 65, which is divided into three routes cleaned by three different trucks. When I was satisfied with my triangulations, I poked my head into the side room to ask Burrafato a question. He was watching TV and sipping a second cup of coffee.

I retreated to my footlocker. Terlizzi was now on the phone with "the borough," his bosses at the Brooklyn headquarters, ordering up an FEL, or front-end loader. "Someone just dumped the contents of the first floor of his house onto the street," he said to me. "Happens all the time." As soon as he got off, the phone rang. It was the cat lady on Fifth Street, complaining yet again that the san men hadn't collected her garbage. A clerk named Scooter handled the call, which meant he held the receiver at arm's length so the entire room could hear the woman's litany of grief. When it was over, he told her soberly, "I'll make sure this information gets to the right people." He hung up, and the assembled burst out laughing. Everyone knew about the cat lady; she owned twenty animals. "It's not against the law to dump your litter box onto your garbage, but it's common courtesy to put it in a bag," Scooter explained.

At last, Burrafato unlocked his sedan and drove me uphill. By now, Murphy and Sullivan were halfway through their route and lightly sweating. The men seemed dour and angry to me, and I was afraid to ask them questions. On foot, I watched and I followed. Soon I realized they seemed sour only because they were concentrating. In constant motion, lifting heavy barrels, they could get hurt if they didn't pay attention. Metal cans banged against their legs; trailer hitches poked from high SUV bumpers. Drivers honked, urging the men to hustle it up, to get their truck out of the way. Double-parked delivery vans blocked their progress. There was also a surprising amount of dog shit near the garbage cans, and many plastic bags were shiny with urine. Had I never noticed this before?

After a few minutes, I began dragging together barrels from neighboring houses to form a group, but the guys didn't want me lifting anything into the truck. "You're gonna be sore tomorrow," Murphy said. He was rounder than Sullivan, and he had a stiff, loping walk, not quite a run. He kept his head mulishly down, his eyes trained on the ground. His palms were thick-skinned and yellow, with deep crevices near the nails. Around the garage, he was known as Daddy. Sullivan had an angular face softened by a narrow strip of beard. His hair was a wiry brown and gray, cut into a mullet. A black belt in karate, he was more agile than Murphy. I found him soft-spoken but intense.

Most people don't think of garbage collection as particularly dangerous work. It may be dirty, boring, and strenuous, but compared to the potential perils of, say, coal mining, the risks in heaving trash seem minor. In fact, the Bureau of Labor Statistics classifies refuse collection as "high-hazard" work, along with logging, fishing, driving a taxicab, and, yes, mining. While the fatality rate for all occupations is 4.7 deaths per 100,000 workers, refuse workers die at a rate of 46 per 100,000. In fact, they're approximately three times more likely to be killed on the job than police officers or firefighters.

Six days a week New York's Strongest—who along with New York's Finest (the cops) and New York's Bravest (its firefighters) constitute the city's essential uniformed services—operate heavy machinery and heave ten thousand pounds in snow and ice, in scorching heat and driving rain. Cars and trucks rip past them on narrow streets. Danger lurks in every sack: sharp metal and broken glass, protruding nails and wire. And then there are the liquids. Three New York City san men have been injured and one killed by acid bursting from hoppers. It takes about a year for a san man's body to become accustomed to lifting five to six tons a day, apportioned into seventy-pound bags. "You feel it in your legs, your back, your shoulders," Murphy told me.

Still, plenty of people want the job. The starting pay is $30,696, with an increase to $48,996 after five years. The health benefits are great, the scavenging superb, and you can retire with a pension after twenty years. With a good winter, one with plenty of snow to plow (in New York, DSNY is responsible for snow removal, which often involves overtime pay), a senior san man can earn $80,000. Thirty thousand applicants sat for the written portion of the city's sanitation test the last time it was offered.

At eight o'clock, truck CN191 turned east onto my block. I saw my downstairs neighbor close our gate and turn with his German shepherd toward the park. "We'll get ten tons today," predicted Sullivan, tossing a black bag into the hopper and cranking the handle. Nine tons had been the norm, but now that the city wasn't recycling plastic and glass, that extra weight landed in his and Murphy's truck.

We moved up the street, about three brownstones at a time, looking for breaks between parked cars. This type of collection was called "house to house." In Manhattan, where high-rises are the norm, san men did "flats," and a truck could pack out after clearing just one or two big buildings. A route in Manhattan might have just three short legs (called ITSAs, though no one remembered why), a route in the lowlands of Brooklyn several dozen.

At last, CN191 parked in front of my building: a brownstone divided into three apartments that shelter six adults, three children, two dogs, two cats, and one fish. (The fish was mine, and it generated very little solid waste: one packet of fish food, I've discovered, lasts three years.) I was nervous. Had we put the barrels—three for putrescible waste, one for metal, and one for paper—in a convenient place? Were the lids off? They were supposed to be on, but they were a pain, and the san men didn't like them. Lids slowed things down. I wondered if someone had dropped a Snapple bottle or a packet of poodle poop into our barrels reserved for paper or metal. Sullivan and Murphy didn't care, but the guys on recycling weren't supposed to collect "contaminated" material, and Burrafato, in theory, could scribble a summons for it. I wondered if my trash was too heavy or too smelly or contained anything identifiably mine. Would Sullivan make some crack about the stained napkins and place mats I was tossing? Would Murphy think it coldhearted to throw out a child's artwork?

Watching for dog shit along the curb, Sullivan rolled one plastic bin to the street and Murphy grabbed two others. They looked heavy—I knew they were about three-quarters full—but the men hoisted them to the hopper's edge without apparent effort. A small plastic grocery sack puffed with refuse, possibly mine, tumbled into the street. My heart almost stopped. Murphy swooped down upon it, tossing the tiny package into the hopper with a flick of his gloved hand. It was over. Nothing untoward had happened. Nobody had said a word.

I suspect that many people feel guilty about the volume of their trash. As I became more educated about garbage, my feelings of shame and guilt grew. There was stuff in my barrel, like those stained linen napkins, for which I'd failed to find further use. When I'd brought this stuff into the house—a new T-shirt, healthful food, a really fun toy—it was live weight, something I was proud to have selected and purchased with my hard-earned money. Now the contents of the bag were dead weight, headed for burial. No wonder we prefer opaque garbage bags. And no wonder that recycling bags, which flaunt our virtue, are often translucent.

Was I being neurotic? What, after all, could Sullivan and Murphy say about me, based on an average week's trash, that couldn't be said about a million others? That I wasted food, made unhealthy snack choices, bought new socks, or had a cold? I knew, after just one day on the job, that san men constantly made judgments about individuals. They determined residents' wealth or poverty by the artifacts they left behind. They appraised real estate by the height of a discarded Christmas tree, measured education level by the newspapers and magazines stacked on the curb. Glancing at the flotsam and jetsam as it tumbled through their hopper, they parsed health status and sexual practices. They knew who had broken up, who had recently given birth, who was cross-dressing.

Sometimes the things one rejects are just as revealing as the things that one keeps, but not always. When sixties radical A. J. Weberman sorted through Bob Dylan's garbage, which he'd snatched from outside Dylan's Greenwich Village brownstone, he found nothing that helped him interpret his hero's cryptic lyrics. Unhappy about this invasion of privacy, Dylan chased Weberman through Village streets and smushed his head to the pavement. The United States Supreme Court ruled in 1988 that the Constitution gives individuals no privacy rights over their garbage, though some state constitutions offer more stringent protection.

Weberman went on to found the National Institute of Garbology, or NIG, and to defend trash trolling as a tool of psychological investigation and character delineation. When he tired of Dylan's garbage, he dove into Neil Simon's (he found bagel scraps, lox, whitefish, and an infestation of ants), Gloria Vanderbilt's (a Valium bottle), Tony Perkins's (a tiny amount of marijuana), Norman Mailer's (betting slips), and antiwar activist Bella Abzug's (proof of investments in companies that made weapons).

Looking through trash often says more about the detective than the discarder. When city officials in Portland, Oregon, decided in 2002 that it was legal to swipe trash in an investigation of a police officer, reporters from the *Willamette Week* decided to dive through the refuse of local officials. What the reporters found most remarkable, after poring through soggy receipts and burnt toast, was how bad the investigation made them feel. "There is something about poking through someone else's garbage that makes you feel dirty, and it's not just the stench and the flies," wrote Chris Lydgate and Nick Budnick. "Scrap by scrap, we are reverse-engineering a grimy portrait of another human being, reconstituting an identity from his discards, probing into stuff that is absolutely, positively none of our damn business."

* * *

At a large apartment building on the corner of Eighth Avenue, Sullivan parked the truck at an angle to the curb. The building's super had heaped long black garbage bags—each a 120-pound sausage—into a four-foot-high mound. It took the team less than two minutes, and a few cranks of the packing blade, to transfer the mound from the street to their truck and crush it all together. When they were done, one bag remained on the sidewalk, its contents gushing through a long tear. "Gotta watch for rats when it's like that," Murphy said, slightly breathless.

"Once a rat ran across my back," Sullivan said. "Whaddaya gonna do?" Maggots, known in the biz as disco rice, were something else. On monsoon days, they floated in garbage pails half full of rainwater. "I won't empty those," Sullivan said.

Before the city's recycling suspension, it was easy for street people to collect deposit bottles for redemption: residents segregated the glass and plastic for them. Now, scavengers tore through everything in the same sacks, heedless of the mess. "It's the homeless," said Sullivan with a shrug. "The super is gonna have to clean this up." A driver with a cell phone to his ear leaned on his horn. Murphy and Sullivan appeared to be deaf.

The ITSAs rolled on and on. I lost track of the street, whether we'd cleaned the left side or the right. Sullivan talked about the seasonal changes in garbage. "In the springtime, there's a lot of yard waste and a lot of construction, because of tax returns. You get more household junk in the spring. You can always tell when an old-timer dies. There's thirty bags and a lot of clothes."

Sullivan continued. "Food waste goes up after Memorial Day and the Fourth of July. You see a lot of barbecue stuff, lots of food waste. And you can always tell when there's a sale on washing machines, usually around Columbus Day."

"People eat different up top," Murphy said, meaning the blocks closer to Prospect Park. "A lot of organic people, fresh stuff. They're more health conscious. There's more cardboard from deliveries; they order those Omaha steaks. People up top read the *New York Times*. They're more educated. In my neighborhood, Dyker Heights, it's all *Daily News* and the *Post*." Though Sullivan thought garbage increased in the summer, with tourists visiting, Murphy thought it went down. "People are away," he said. "In October, you get a lot of rugs and couches." Harvest season.

The way residents treated their garbage said a lot about them, in the san man's world. "In the neighborhood where I live, the garbage is boxed and gift-wrapped," Joey Calvacca had bragged to me. For the last seventeen years, Calvacca had been working in the Brooklyn North 5, in East New York. Though he'd long ago moved from the city to Long Island's North Shore, he still spoke in the dialect of *The Sopranos,* eliding all *r*'s. "But where I work, it's a mess. People don't use bags. There's maggots, rats, roaches. The smell will make you sick. I've gotten stuck with needles."

"And what about your garbage?" I asked.

"It's normal garbage," he said, shrugging.

Good and *bad* referred to garbage content as well as garbage style. Good garbage, the san men taught me, was garbage worth saving. They called it mongo. The sanitation garage was brimming with it: a microwave, a television, chairs, tables. "Some neighborhoods in Queens, the lawn mower is out of gas and they throw it out," Calvacca said. "They throw out a VCR when it needs a two-dollar belt. We throw it in the side of the truck to bring home." Silk blouses and designer skirts billowed from the trash of upscale buildings. Tools and toys, books and bric-a-brac were there for the taking. Officially, mongo didn't exist. Sanitation workers weren't allowed to keep stuff they found on the curb. But everyone did, and no one complained.

The truck was about two-thirds full now. Inside, brown gunk dripped off the packing blade into a nest of ratty clothing. Rounding the corner onto Seventh Avenue, Sullivan and Murphy pulled over to gulp from water bottles and wipe the sweat from their foreheads. I felt chilly in a rain jacket over a fleece pullover. Their cotton shirts had bibs of sweat. On 95-degree days, Sullivan said, he went through three T-shirts in one shift. In the rain, he didn't even bother with a slicker. "You're soaked from the inside anyway, water running down your neck," said Sullivan. "It's awful."

I asked how close they were to finishing today. "We'll do it all in three and a half hours," said Sullivan. "That's without a coffee break or lunch."

"Why do you work so fast?"

"To get it over with," said Murphy.

That didn't exactly explain the panic to finish early. San men couldn't go home when their job was done; they had to stay in the garage until their shift ended, at 2:00 p.m. The men would pass the time eating lunch, watching videos or TV in the break room, playing cards, and working out on exercise equipment rescued from the jaws of the hopper. "We used to have a pool table, but it wore out," Sullivan said. Now the men napped on white leather couches, relics from another era. (From garage to garage, break room decor varied enormously, constrained by the availability of local mongo, the super's aesthetic sensibilities, and the culture of the particular garage. Now and then, a call from "downtown" resulted in a clean sweep, and all the bad paintings, ceramic kitsch, macramé wall hangings, tin signs, plastic flowers, hula hoops, and velvet Elvises went into the garbageman's garbage pail.)

"The time passes quickly," said Sullivan. "You're coming down from a big high afterward. It's like an athletic event." He screwed the cap onto his water bottle. "I figure it's the length of a marathon, every day. You just try to get through it. You can't think about it. It's a state of mind."

In 1993, Italo Calvino published an essay about his daily transfer of trash from the kitchen's small container to a larger container, called a *poubelle,* on the street. "[T]hrough this daily gesture I confirm the need to separate myself from a part of what was once mine, the slough or chrysalis or squeezed lemon of living, so that its substance might remain, so that tomorrow I can identify completely (without residues) with what I am and have." He equated his satisfaction with tossing things away to his satisfaction with defecation, "the sensation at least for a moment that my body contains nothing but myself."

I felt a kinship with Calvino, for I was obsessed with throwing things away. Transferring objects—whether food scraps, the daily newspaper, or a lamp—from my house to the street made me feel lighter and cleaner, peaceful even. My apartment wasn't large, and so everything I subtracted gave me more of what I craved: emptiness.

Eventually, Calvino came to realize that so long as he was contributing to the municipality's waste heap, he knew he was alive. To toss garbage, in his view, was to know that one was *not* garbage: the act confirmed that "for one more day I have been a producer of detritus and not detritus myself." Riffing on death and identity, Calvino referred to the men who collected his garbage as "heralds of a possible salvation beyond the destruction inherent in all production and consumption, liberators from the weight of time's detritus, ponderous dark angels of lightness and clarity." In a similar vein, Ivan Klima, in his novel *Love and Garbage,* noted that street sweepers regard themselves as "healers of a world in danger of choking." My san men, while not obviously self-reflective, knew exactly how the public viewed what they did: "People think there's a garbage fairy," one worker told me. "You put your trash on the curb, and then *pffft,* it's gone. They don't have a clue."

When their truck was full, at around ten-thirty, Murphy dropped Sullivan at the garage, then rumbled over the Gowanus and pulled into the courtyard of the IESI transfer station, a white-painted brick building at the corner of Bush and Court Streets, in Red Hook. The drill here was simple: weigh your truck, then pull around to the tipping floor, back in, and pull the lever to dump. If the men had loaded their truck properly, the ejected garbage would extend six to eight feet in a supercompressed bolus before dropping to the ground. Garbage that simply spilled out was poorly packed or indicated that the truck hadn't been full. The quality of the dump was known as "the turd factor." According to one designer of packer trucks, "The driver can learn from experience by observing the turd factor and know just how much trash he can put in the truck per trip. If he gets a good turd on every trip to the landfill, that's a good day." Judging by the conformation of today's load, Murphy and Sullivan had done well. The morning's labors—twenty thousand pounds collected in less than four hours—now lay in a heap, indistinguishable from the heaps dumped before or after. Without a backward glance at what he'd deposited, Murphy put the truck in forward gear.

Integrated Environmental Services Incorporated was founded in 1995 and had grown by acquisitions, gobbling sometimes as many as a hundred companies a year. It bought this facility from Waste Management in 1999. By 2003, IESI was the tenth-largest solid waste company in the nation. In New York it was the third largest, after Waste Management and Allied Waste.

I pedaled down to the Court Street station a few days after going out with Sullivan and Murphy, hoping to speak with the plant manager. The doors on the bays were closed, and no one was about. I saw no trucks, and I smelled no garbage. If you didn't know what went on inside this building, you wouldn't, on a cold and slow autumn day, be able to guess. I rode around the neighborhood, noting that it was zoned for industry. Then I turned a corner, from Bush onto—just coincidentally—Clinton, and now the multiple towers of the Red Hook housing projects, home to eleven thousand mostly low-income residents, rose before me.

Garbage follows a strict class topography. It concentrates on the margins, and it tumbles downhill to settle in places of least resistance, among the poor and the disenfranchised. The Gold Coast of Manhattan's Upper East Side produces far more waste than the neighborhood of Williamsburg, in Brooklyn, but city officials have never tried to site an incinerator on Park Avenue, as they did in Williamsburg. Similarly, landfills have no place within the city limits of Grosse Point, Beverly Hills, or Palm Beach. Across the nation and around the world, trash is dumped, metaphorically, upon trash.

Of course, there are, in this era of modern landfills, plenty of communities that say yes to trash, even to trash generated far, far away. The inducement is cash—or its in-kind equivalent. Tullytown, Pennsylvania, population 2,200, has raked in $48 million over the last fifteen years in exchange for burying fifteen million tons of trash, mostly from New York and New Jersey. Literally, this is the town that garbage built: waste paid for its town hall, half the police force, the new fire truck, the marine rescue boat, playground, trees, sidewalks, lampposts, and fireworks. (Towns that agree to host dumps invariably get free garbage pickup, too.) The town of Taylor, Pennsylvania, receives a minimum of $1.5 million a year from the Alliance landfill for hosting a five-thousand-ton-per-day landfill, in addition to a one-time lump sum of $900,000, which paid for a new library and a senior center.

Lee County, one of South Carolina's poorest, receives a fifth of its annual budget from Allied Waste, which pays $1.2 million a year to dump there. Sumpter Township, in Michigan, turns a fraction of Toronto's waste into nearly half its annual income. In 2003, Waste Management paid Michigan's Lenox Township nearly $1.8 million, which it used to improve a park, buy two EKG machines, and acquire two thermal-imaging cameras for the fire department. Charles City County, in Virginia, lacks a supermarket, drugstore, and bank. But after the Chambers Development Company built a supersized landfill there, the county cut property taxes (Chambers pays 30 percent of the county's operating budget) and started to build schools. In Canton Township, Michigan, the Auk Hills landfill contributed $13 million to build the town's Summit on the Park community center. (These deal sweeteners aren't unique to trash and tiny towns: before New York City could build a sewage treatment plant on Manhattan's far Upper West Side, it promised the community a twenty-eight-acre park, complete with soccer field, indoor and outdoor swimming pools, and an ice-skating rink—all sitting smack-dab atop the settling tanks and sludge thickeners.)

Giant waste companies don't mind paying host fees: they help smooth over community opposition and legal hassles. Christopher White, president of Mid-American Waste Systems, explained the historical setting of host fees to a *Forbes* reporter: "It's something the utility companies and the railroads have done for years." In the dozens of tiny towns that were exploited, then polluted and abandoned by King Coal before being forced to contemplate megadumps in their scarred backyards, this type of justification, made by an absentee power lord, probably isn't all that reassuring.

"People get very rich very fast if they're willing to impose on a poor community that can't fight back," Al Wurth, a political scientist at Lehigh College, in Bethlehem, Pennsylvania, told me. "There are enormous incentives for certain groups to do this. They're not thinking about the effect of stuff three generations from now. They'll be gone. But the stuff lingers on." It is an especially raw deal for neighboring towns that aren't getting new ball fields and Fourth of July fireworks. They get all the truck traffic, the air and water pollution, the birds, the stench, and the degraded property values, but all the host benefits lie just over the county line.

A decade before Fresh Kills was slated to close, New York City officials went shopping for a new place to dump. One destination under consideration was West Virginia's McDowell County, near the state's southern border. Facing acute unemployment and underdevelopment, the town of Welch, the county seat, saw no better economic alternative than to build a landfill in a bowl-shaped hollow at the end of Lower Shannon Branch, a dirt road that winds for six miles through hill country.

In exchange for accepting 300,000 tons of waste a month, most of it from New York City, Welch would receive an $8 million fee from the development company, 367 jobs, and one wastewater treatment plant, a novelty for a county that, by dumping raw sewage into its creeks, had been in violation of the Clean Water Act since 1972. Only a handful of people had questions about the project, but just as the contract was about to be signed, a protest movement materialized. Much was made of the waste's provenance: accepting garbage from New York and New Jersey, the landfill would surely be tainted with AIDS and by medical waste, it would be run by the mob, and "cocktailed" with toxic and nuclear dregs. (Homegrown trash, presumably, didn't even smell.) The plan was ultimately defeated by economics, despite a referendum in favor of the dump. In 2004, the landfill's developers presented a reworked proposal for McDowell to the state legislature. After all, the county was still in desperate financial straits, its creeks still flowed with sewage, and New York was still producing waste.

Across the nation, environmental justice groups have sprung up to fight the siting of transfer stations, landfills, wastewater treatment plants, and other polluting industries within low-income neighborhoods and communities of color. According to a 1987 study conducted by the United Church of Christ's Commission for Racial Justice, three out of every five African Americans or Hispanic Americans live in communities with one or more unregulated toxic-waste site. These environmental justice groups track cancer and asthma clusters, educate their constituents, and work to clean up hazardous waste.

In my travels with trash I learned that more than two-thirds of New York City's residential and commercial waste flows through transfer stations in just two neighborhoods: the Bronx's Hunts Point and Brooklyn's Greenpoint-Williamsburg. (Altogether, the city has sixty-two land-based transfer stations, not one of which is located in Manhattan.) It isn't just garbage that irritates the stations' neighbors. Six days a week, twenty-four hours a day, ten-ton packer trucks roll in with their deliveries—at some stations, more than a thousand of them a day. Altogether, they travel a total of forty thousand miles a day, trailed by a diesel plume of particulate matter. According to Inform, an independent research firm that examines how business practices affect the environment and human health, packer trucks account for only 0.06 percent of the vehicles on US roads, but they consume more fuel annually—and discharge more pollution—than any vehicles other than tractor-trailers and transit buses. Why do garbage trucks have such a heavy impact? Because they cover twice as many miles per year as the typical heavy-duty single-unit truck, and they travel less than three miles on a gallon of gas.

Greenpoint, home to sixteen waste transfer stations processing about a third of the city's garbage, has the highest concentration of airborne lead in New York City, and the second-highest rate of asthma. Epidemiologists link the disease with particulate matter smaller than two microns, the stuff that spews from the stream of packer trucks bringing garbage in, and from the tractor-trailer trucks that idle in a queue, waiting to haul it away.

Since Fresh Kills closed, almost all of the city's waste is trucked from transfer stations to out-of-state landfills and incinerators. According to Keith Kloor, reporting for *City Limits,* it takes about 450 tractor-trailer trucks to complete this task each day, burning roughly 33,700 gallons of diesel fuel. The combined round trips add up to 135,000 miles. An additional 150 packer trucks, carrying about fifteen hundred tons of waste a day, make shorter trips to three incinerators in New Jersey and Long Island. The trucks wear down city streets and outlying highways, and their emissions of carbon dioxide, sulfur dioxide, chlorofluorocarbons, and other pollutants contribute to elevated asthma and cancer rates, acid rain, ozone depletion, and global warming.

The cost of shuttling city garbage around the boroughs and out of state is not cheap. In tolls alone, the city spent $2.25 million in 2002. Trucking and tipping fees cost another $248 million. Including the hiring of three hundred additional drivers to relay full trucks to transfer stations, the city spent $257 to dispose of each ton of trash in 2002, a 40 percent increase over the 1996 cost.

My garbage was now in private hands. To get a look at it, I had to call Mickey Flood, the CEO of IESI, in Fort Worth, Texas, and then Ed Apuzzi, the company's vice president for business development and legal affairs in the Northeast region, who decided we should meet at the transfer station on Election Day, when DSNY wasn't delivering garbage (though commercial waste continued to pour in). At the appointed hour, I stood at the building's corner and waited for Apuzzi to show. The sidewalk was litter free but greasy. A truck had damaged the corrugated metal fence across the street, and there was a deep pothole on the corner filled with opaque gray liquid. The building had recently been painted white with blue trim. Under the company logo—a pine tree—was a phone number to call with any complaints.

Casually, as if I weren't really spying, I glanced inside the transfer station. At first, I couldn't tell what I was looking at. Like a Hollywood soundstage, the walls, floor, and ceiling were painted black, and there were large floodlights mounted on tracks overhead. But there weren't many of them, and they shed only a dim light on the hilly mosaic of garbage that covered half the floor. Higher up, they illuminated what I at first took for dust motes but realized, when I got a little closer, were droplets of a powerful perfume, which shot from nozzles near the ceiling. The smell was sweetly antiseptic. As my eyes adjusted to the light, I made out large black bags of garbage, small supermarket sacks of garbage, one of which could have been mine, some bulk metal pushed off to one side, a rotted board, chair cushions, a ketchup bottle. But still, the upper contours of the space were indeterminate. I could have been in a planetarium.

At first the jumble of goods, some ten feet high, appeared homogeneous to me: it was just a lot of garbage—dirty, ragged, bagged, loose. But to the practiced eye of a fanatic recycler or a Mexican *pepenador*, a professional trash picker, the pile was actually heterogeneous. It contained metals and textiles, wood and glass—commodities with value. Save for the preponderance of plastic, it comprised almost the same materials found in a nineteenth-century ragpicker's shanty: bones, broken dishes, rags, bits of furniture, cinders, old tin, useless lamps, decaying vegetables, ribbons, cloth, legless chairs, and carrion.

Back in the day, all of those items would have found another use. Today, they were prodded into a rough pile by a worker in a front-end loader and spilled into a tractor-trailer parked along the far wall of the tipping floor. Using the backside of its bucket, the loader awkwardly patted the reeking mass into one solid rectangular cube. The driver tucked a tarp over the garbage and, with a roar of the engine, was gone.

While I waited for Apuzzi, I made small talk with Frank Morgante, the site manager. I asked him if neighbors complained about the station.

"They walk by here and they give us looks," he said. "They look at us like we're garbage. I want to say to them, 'You want to solve the garbage problem? Stop eating. Stop living. Then we won't have any more garbage.'"

A middle-aged man walked by, and I asked him what it was like living near a transfer station.

"IESI is *not* a good neighbor," he said. "The place smells and it's overrun with rats."

"We have an exterminator every two weeks!" Morgante interjected.

"That just sends them up there," the man shouted, indicating his home in the Red Hook housing projects.

Earlier, Morgante had told me that rats tumbled out of the trucks—they weren't living at his transfer station. Now he told the neighbor that the trash didn't sit in the transfer station. It came in and it went out. In. And out. He repeated it brightly. The neighbor didn't care about in and out. He cared about the continual presence of garbage. He cared about its cumulative physical impact.

"This place has given me asthma," he said.

"You probably had asthma before we ever worked here," Morgante said. They were getting a little loud. The neighbor waved his arm as if to ward off Morgante's retort and turned to leave. "You know why the garbage is here?" he asked. "It's because we're poor."

"You know what?" Morgante said to his back. "I'm poor, too, and I don't live that far from here."

Apuzzi finally appeared. He was clean shaven and tanned, with a neat haircut. He seemed ill at ease here. He declined to wear the orange vest and hard hat that Morgante had forced upon me before I made the ten-foot walk from the bay door, over the knee-high tide of garbage, to an open stairway that led to a small office. Frowning in his dress shirt and polished brown shoes, Apuzzi picked his way over a sofa cushion, across the slippery frame of a foldout bed, and in between two black garbage bags. A sheen of brown muck coated the floor.

The office, which smelled slightly garbagey, contained a cheap L-shaped desk with a computer, a small meeting table, and several ceiling-mounted security monitors. The room had no street windows, but it did have an interior window that overlooked the tipping floor, and that's what I wanted to see.

At 11:00 a.m., the trash was halfway up to the horizontal yellow line on the push wall. The front-end loader, with its six-yard bucket, was filling a truck. I asked Apuzzi why everything was painted black. "I don't know," he said. He seemed as puzzled as I was.

I asked him about his background. I imagined that like many in the trash business, he was a guy whose career had probably started out promisingly enough in another field but had then taken a sudden turn and rolled downhill. "I'm an attorney," he said. "I worked as a litigator in Manhattan and then Princeton until IESI bought my family's waste collection business." Now his boss was Mickey Flood.

"Trucks dump here until about eleven p.m.," Apuzzi said, gazing down on the trash. "The floor has to be clean by midnight—empty of garbage and washed. Then the garbage starts coming in again." I watched as a kid from the projects zipped around in a small forklift, picking bulk metal objects from the trash heap—a stroller, a desk, a swing set frame. He piled this stuff in the station's adjacent empty bay. Metal is heavy, and IESI didn't want to pay to tip it in someone else's landfill. The company could sell it for scrap. "The garbage always sits less than one day," Apuzzi continued. "On Sunday we're empty."

I asked how many trucks came in each day. "City trucks bring about eight tons each and commercial trucks bring thirteen. You can do the math." I divided the station's permitted 745 tons by 21 tons, the amount in one commercial and one residential truck, and got approximately 75 full trucks entering each day. The tractor-trailer trucks held 20 tons, so that was an additional 37 trucks leaving. They delivered the waste to two landfills IESI owned in Pennsylvania or—if those landfills had met their daily permitted tonnages—to two or three others owned by competitors.

"And you always get your seven hundred forty-five tons?"

"We always make our quota," Apuzzi said. "When we're full, we let the Brooklyn garage know, and they divert trucks to other transfer stations."

The number of truck trips out of transfer stations—450 tractor-trailers a day—was a flashpoint for garbage activists. Environmental and local advocacy groups wanted the city to reopen its marine transfer stations—there was at least one for each borough, excluding Staten Island—and barge the garbage from neighborhoods. The packer trucks would still drive in, but tractor-trailers would be eliminated from the equation.

"If the city opens its marine transfer stations, we'll do just commercial," Apuzzi said. His permit was good for 745 tons—it didn't matter where it came from. But he didn't think garbage barges were coming anytime soon. "It will cost the city more at the marine transfer stations because containerization is significantly more expensive. Retrofitting the stations will cost hundreds of millions of dollars." (Apuzzi was right, but just for a year: after announcing it would fix up the transfer stations, the city backed down from the plan, citing its expense, then recommitted.)

The city now paid IESI an average of sixty-five dollars a ton to tip residential waste. "What we pay to tip at landfills fluctuates," Apuzzi said. "If the distance to the landfill is long, it costs us more to get there, but tipping fees are lower."

"Where do you usually go?"

"Bethlehem, Pennsylvania. We own a landfill there."

"I'd like to visit it, see what happens to my trash next."

Apuzzi narrowed his eyes and wrote something on his legal pad. "I'll see what I can do," he said, not sounding confident.

I asked if there was community opposition to the transfer station. "Historically, yes. But since IESI has been here, we run an efficient and environmentally sound operation. We run with the doors closed. We don't allow garbage to spew out. We use a perfume neutralizer. We have little traffic. We've got a ventilation system with a scrubber on the exhaust. Over the years, community opposition has dwindled." He looked at his watch.

"Let me ask one more thing," I said. "Do you think we have a garbage problem?"

"That's a good question," Apuzzi said, sitting back in his chair. He was quiet for a moment, then, "No, I don't think we do. We have plenty of room for it. It would be nice to have a landfill within the city boundaries. But I don't think Fresh Kills is going to reopen soon. That place was an environmental nightmare." I smiled to myself. Just as individuals imagined their trash was better than the next guy's, so did dump owners think their operations were better than the next dump owner's.

"I don't know a thing about Fresh Kills," I told Apuzzi. "I'm still waiting for someone to let me in there."

With that, our interview was suddenly over, and Apuzzi ushered me from the office. I never saw IESI's vice president for business development and legal affairs in the Northeast region again, despite repeated attempts to meet up with him. In any case, my attention would soon turn from transfer stations to landfills. Although it was no longer the final destination of my garbage, the Fresh Kills Sanitary Landfill was the K2 of trash heaps, and I was determined to make an assault on its closed and forbidden slopes.

Chapter Two

Amphibious Assault

New York City was not unusual in shunting quantities of noxious waste to its backyard. Every American city, up until about the middle of the twentieth century, dumped its rejects on nearby scraps of low-value land—usually in swamps. In 1879, a minister described the situation in New Orleans to the American Public Health Association:

> *Thither were brought the dead dogs and cats, the kitchen garbage and the like, and duly dumped. This festering, rotten mess was picked over by ragpickers and wallowed over by pigs, pigs and humans contesting for a living from it, and as the heaps increased, the odors increased also, and the mass lay corrupting under a tropical sun, dispersing the pestilential fumes where the winds carried them.*

When swamps grew scarce, holes were excavated in dry land and garbage was tumbled in. Sometimes the trash was burned to reduce its volume, a process that created billows of black smoke and toxic fumes. In 1937, Jean Vincenz, the commissioner of public works for Fresno, California, after traveling the dumps of the nation, built in his hometown the country's first "sanitary landfill" (England had a sanitary landfill, too, but it never accepted household waste). Every day, Vincenz carefully positioned and compacted the city's waste. Then, to keep down vermin, birds, and odors, he buried it with soil that he'd dug out to make room for the next day's haul. The sanitary landfill idea began to catch on during World War II, when the US military adopted Vincenz's methods. By 1945, according to the Garbage Project's William Rathje, one hundred cities had joined the bandwagon, and by the fifties, the sanitary method was in full flower.

While Vincenz's approach made America's trash heaps look cleaner, their subterranean aspects were anything but. A blanket of dirt didn't protect groundwater from contaminated moisture, called leachate, seeping through the garbage, and it didn't control or capture leaking landfill gases, which are toxic. It wasn't until the 1980s and '90s that the Environmental Protection Agency, through its Resource Conservation and Recovery Act, began to protect human health and the environment against these discharges by requiring leachate- and methane-collection systems. In 1991, the agency gave landfills six years to modernize or close. But complicated liners made of plastic and clay, and gas-collection-and-monitoring systems were expensive, so most dumps shut down. In 1988, there were nearly 8,000 landfills across the country; in 1999, there were 2,314; and by 2002, there were only 1,767.

(A historical footnote: Jean Vincenz's Fresno landfill had, by the time the RCRA rules came out, long been shuttered. But it received a flicker of attention on August 27, 2001, when Secretary of the Interior Gale Norton nominated it as a National Historic Landmark for its importance in the nation's history of civil engineering. The fuss lasted less than twenty-four hours. The very next day, Norton rescinded her offer: the landfill, she'd just learned, was an EPA Superfund site.)

As illegal dumps closed, fewer but larger regional landfills were inaugurated, usually in rural areas with small populations. Meanwhile, Americans continued to generate more and more trash. The garbage business—concentrated in the hands of a few major corporations—blossomed. By 2001, it was a $57-billion-a-year industry. The economics of megafills covering several hundred acres was irresistible. They were, in relative terms, far cheaper to build than little landfills, even when the millions paid to control pollution and fight community activists were factored in. Because megafills were built in small sections, or cells, with revenue coming in as each section filled, their construction costs could be spread out over many years. In this way, a large landfill could reap gross profits of more than 50 percent.

With urban dumps capped, more and more waste began to cross state borders. Dump owners adopted a "smoke 'em if you got 'em" attitude, working to fill their sites as quickly as possible—before federal regulations could curtail interstate trade in garbage, before recycling could claim an even larger percentage of the waste stream, before environmental regulations tightened up and raised their cost of operations. Just as road builders invited more traffic by adding lanes to highways, so did enormous landfills invite a wanton disregard for waste reduction. If there was plenty of room out there, what was the incentive to conserve space?

In 2002, thirty-two states (reporting to *BioCycle* magazine) imported garbage from other states, while twenty-four states exported garbage to other states (some states did both). Pennsylvania led the importing pack, accepting ten million tons in 2002. In second place came Illinois, followed by Virginia, which shipped its own hazardous waste primarily to dumps in Ohio and New York State. The second-largest exporter was New Jersey and the first was New York, which had contracts to dump its garbage at thirty-seven landfills in New Jersey, New York, and Pennsylvania (with permits to dump in Ohio, Virginia, and South Carolina if need be), in addition to four incinerators in three states.

Before it closed in 2001, Fresh Kills had been the largest city landfill in the country, indeed, in the entire world. That honor is now held by the Puente Hills Landfill, in Whittier, California, which has a permit to accept twelve thousand tons of trash a day from the surrounding counties. Not to belittle Whittier, but that's just half the daily tonnage that was dumped into Fresh Kills, which sprawled over three thousand acres on the western side of Staten Island.

I'd requested permission from the city's Department of Sanitation to visit Fresh Kills because it played such an important role in New York's garbage history, but my application seemed to be in limbo (much like my request to visit IESI's Bethlehem landfill). I asked Sanitation one more time for a tour of Fresh Kills, then took matters into my own hands. Studying a map of Staten Island, I made out a green-tinged area just north of the dump. In fact, it seemed to be connected to Fresh Kills by a tidal creek. I'd never heard of the William T. Davis Wildlife Refuge, but it seemed like a good place to start.

"Do you have a boat that I might be able to rent?" I asked the receptionist at the refuge. I implied that I was interested in the area's native grasses. No, she said, the refuge didn't keep rental boats, but the resident naturalist, Carl Alderson, knew a lot about the grasses. Without waiting for my response, she patched me through.

I was getting a bit off track, but I liked talking to Alderson. He was refreshingly warm and friendly. A salt marsh ecologist, he had restored wetlands in the Arthur Kill, which separated Staten Island from New Jersey, after a big Exxon oil spill in 1990. (The word *kill* is itself a sort of historical debris. Dutch for *river,* it remains, despite the conquest of the English, scattered throughout the New York area, from Fresh Kills to the upstate towns of Fishkill, Peekskill, and beyond.) Alderson was studying salt grasses on New Jersey's Rahway River, and he'd completed a three-acre marsh restoration inside Fresh Kills itself. My ears pricked at this news: now we were getting somewhere.

We talked a little about the effect of garbage on the environment, and then Alderson revealed that he had four canoes, a kayak, and a Boston Whaler at his disposal. He asked if I wanted to give him a hand counting spartina grass stems in a week or two at the Raritan site. Sure, I said, trying to temper my enthusiasm. To get to New Jersey, I figured, we'd have to boat right past the landfill.

We set a date for fieldwork, but early that morning it rained. We rescheduled, and once again got rained out. Weeks passed, the weather didn't improve, winter closed in. Alderson's spartina grass had wilted, and I knew he couldn't count his stems until next spring. I was about to give up on seeing Fresh Kills from the water when, one weekend in the middle of December, the telephone rang.

"Come on," said Alderson. "Let's paddle around the landfill."

"Okay," I said, fist raised in silent triumph.

It was 25 degrees and the sky was leaden by the time I met Alderson at the refuge office. With his square chin and snub nose, and in his dark fleece jacket and jeans, he looked a little like George W. Bush, but taller. He had longer hair, too, and a different set of politics.

Together we strapped a three-person kayak onto Alderson's battered Toyota, then drove to our put-in spot, on Travis Road. We manhandled the boat down a rocky grade and into a creek that was just a few inches wider than the kayak. As we adjusted the foot pegs it began to snow, but neither of us mentioned this development. The slate gray mud stretched out for yards on either side of us. I heard the rhythmic *thwack* of a hammer to the north. New houses were going up just outside the refuge, while old ones leaked raw sewage into the creek. "I try to avoid touching this water," Alderson said.

I didn't know what to make of the refuge, which had been the first wildlife sanctuary in New York City. Our put-in, high on Main Creek, lay amid a twelve-square-mile complex of derelict brownfields and thriving industry. There was a paper mill to the southwest, the landfill to the east, and a checkerboard of oil tanks to the west across the Arthur Kill, which carried more boat traffic annually than the Panama Canal.

Hemmed and hampered by this built environment, the William T. Davis Wildlife Refuge provides shelter for hundreds of native plant and animal species. Founded in 1928 with a land purchase by the Audubon Society and named after a Staten Island naturalist, its 260 acres contain salt meadow, low marsh, forested uplands, rock outcrops, a swamp forest, spring-fed ponds, and alluvial dunes. Those dunes, which protrude like whale backs above the marsh, are composed of sandy deposits that were swept up by strong currents or prevailing westerly winds from the bottom of Lake Hackensack when its water level plummeted some ten thousand years ago. The lake itself, which once covered the western shore of Staten Island and the eastern shore of New Jersey, was formed by the terminal moraine of the retreating Wisconsin glacier. When the lake drained (possibly due to a breach in a natural earthen dam), its rushing waters carved the narrow channel known today as the Arthur Kill, into which Main Creek flows.

We stroked downstream through black water flanked by mudflats and brown grasses. A raft of mallards pivoted in the rising wind. Then suddenly, as we rounded a bend, the mound of Fresh Kills' Section 6/7 rose before us. (The garbage at the landfill was heaped into four enormous mounds, called sections.) I had expected something massive—the size of the dump impresses every visitor, and few media accounts fail to mention that astronauts can see Fresh Kills from low Earth orbit. But from the water, Section 6/7 was just a steep hill jutting maybe a hundred feet above the deck of our bow, its knee-high brown grasses undulating in the breeze. A few dump trucks, burdened not with garbage but with dirt, trundled up switchbacks, black dots on the landscape. The air smelled lightly of sulfides. Crushed plastic bottles and old tires festooned the creek sides.

With our fingers growing numb, we shipped our paddles and shoved our hands inside our jackets. I turned to look north, toward the refuge, and was struck by the beauty of this place, the alacrity with which the marsh turned into scrub forest—white oak, red maple, sweet gum, and black willow. Before the city began dumping here, this patchwork of marsh, meadow, and forest was the norm. "The landfill supplanted one of the largest tidal areas on the East Coast," Alderson said. "The salt marshes of Fresh Kills, Jamaica Bay, and the Hackensack Meadowlands were unrivaled in their abundance and diversity."

Fifty-three years ago, the writer Joseph Mitchell spent some time in the marshes with Happy Zimmer, a shellfish protector for the Bureau of Marine Fisheries. Zimmer described the marshes and their uplands as a busy and bounteous place. Locals hunted for mushrooms in the autumn, dandelion sprouts in the spring, mud shrimp, herbs, and wildflowers in the summer. Wild berries and grapes went into jams; watercress from freshwater tributaries made salads. In 1929, the Crystal Water Company began bottling and selling springwaters that meandered down toward the sea. Zimmer noted great numbers of pheasants, crows, marsh hawks, black snakes, muskrats, opossums, rabbits, rats, and field mice.

By 1951, when *The New Yorker* published Mitchell's story about Zimmer, the city had already filled in five of the "once lovely" clay pit ponds at Greenridge, on the dump's southern flank. "The marshes are doomed," Mitchell wrote. "The city has begun to dump garbage on them. It has already filled hundreds of acres with garbage. Eventually, it will fill in the whole area, and then the Department of Parks will undoubtedly build some proper parks out there, and put in some concrete highways and scatter some concrete benches about."

The current nudged us toward shore, and Alderson raised his binoculars. "Ring-billed gull," he said. "Canada goose. Great blue heron. Geotextile lining." He aimed his paddle at a scarf of exotic black lying against the brown of Section 6/7. The high-tech tarp was part of Fresh Kills' final closure plan, a barrier that was supposed to keep rain from the garbage and provide footing for future soil and vegetation.

When the EPA began examining landfill impacts in the 1980s, it adopted the "dry tomb" philosophy of landfill construction, which focused on isolating garbage from its immediate surroundings. New landfills, the agency said, would have liners that protected groundwater from leaking garbage juice; collection pipes to funnel this juice into treatment plants; methane-collection pipes to vacuum the gases created by biodegrading organic material; and, when it was time to close the landfill, some sort of plastic layer that would act like an umbrella and keep rainwater from percolating through the waste. Dry garbage, went the argument, was inert, quiet, and calm. Wet garbage, engineers knew, would generate leachate for thousands of years: the dumps of the Roman Empire, more than two thousand years old, are still leaching today.

But there's one problem with dry-tomb landfills: plastic covers and plastic liners break. It is widely acknowledged, including by the EPA, that even the best plastic will ultimately leak, and well before the waste it contains ceases to threaten the environment. How long does waste pose a threat? According to G. Fred Lee, an environmental engineer who's devoted his entire professional life to the study of landfills, "For as long as the landfill exists."

And so in recent years, a new philosophy of waste has edged its way into civil and biological engineering circles. The so-called bioreactor method is just the opposite of the dry tomb, calling for leachate to be collected and then repeatedly injected back into the garbage. All this moisture accelerates decomposition, so that bacteria feeding off waste produce more gas more quickly. After the fermentation of waste has stopped, the dump contents are rinsed with fresh water, and the toxic runoff is collected and treated before final discharge. Because the garbage shrinks while it decomposes, the landfill settles and stabilizes faster (while monitors are still keeping an eye on things, it is to be hoped). Landfill managers like the idea of this method because they don't have to collect and continually treat their leachate. But because there is a lot more of it cycling through their dump, there is more liquid to potentially leak. Leak-detection liners do exist, but they are very expensive and, said Lee, "No one uses them." While university scientists in biological engineering departments are busily running wet-tomb bioreactor models in bins outfitted with Plexiglas windows, the

EPA is monitoring long-term experiments on four actual bioreactor landfills in California, North Carolina, and Virginia.

"Another problem with bioreactors," said Lee, who actually favors the wet method, "is that most household garbage is sealed inside plastic bags, which get compressed by big machines but not always ripped open." And if the garbage isn't exposed to moisture, it might as well be sitting in a dry tomb. Shredding every bag before it's buried would solve the problem, but this would add about a dollar a ton to tipping fees, said Lee. The job may be too messy even for a dump.

Fresh Kills is neither a dry tomb nor a wet bioreactor. Like any landfill, it produces leachate, a noxious stew of household toxics, such as battery acid, nail polish remover, pesticides, and paint, combined with liquid versions of rotting food, pet feces, medical waste, and diapers. An analysis of free-flowing leachate sampled from landfills of the Hackensack Meadowlands turned up oil and grease, cyanide, arsenic, cadmium, chromium, copper, lead, nickel, silver, mercury, and zinc. To capture the leachate flow at Fresh Kills, which shares this pestilential profile, engineers encircled the uncapped garbage mounds with leachate walls dug seventy feet deeper than the lowest layer of garbage. Comprising perforated pipes, the walls funnel garbage juice to the dump's private water treatment plant, where it runs through an intensive detoxification program before being discharged into the Arthur Kill. The leachate walls, in place only since 1996, capture much of the three million gallons that flowed daily before the landfill was capped, but the laws of hydrogeology will not be denied. Ancient streams and channels still run through the former clay pits (remember, they aren't lined), and twice a day tides still flush escaping leachate into New York Harbor.

Liquefied parsley stems and mucus-filled tissues are gross, but leachate from residential garbage has some far nastier characteristics. In addition to pathogens from organic waste, it also picks up metals and acids, motor oil, and solvents from ordinary compounds used in the home. According to Garbage Project studies, about 1 percent by weight of all household garbage could be considered hazardous by EPA standards. Nail polish, for example, contains several chemicals listed by the EPA as hazardous. According to researchers at Texas A&M University, the leachate produced inside Subtitle D landfills, which contain only municipal solid waste, and Subtitle C landfills, which contain hazardous waste, is chemically identical.

I wasn't throwing out drums of rat poison or anything else that seemed chemically risky, but since recycling had been suspended in New York, I tossed out plenty of plastic containers. I was surprised to learn that most household cleaners and shampoos come in bottles that contain the industrial organic chemical dicyclohexyl phthalate, a plasticizer that's suspected of disrupting the endocrine system and harming the liver.

State-of-the-art landfills have composite liners that conduct leachate to a plant where it is treated, then discharged to a local waterway. But, again, even the most sophisticated liners eventually leak. Geomembranes are eaten away by common household chemicals, stuff like mothballs, vinegar, and ethyl alcohol. (I was guilty of discarding the first two items, but not the third: my booze bottles were always empty when I threw them out.) And then there's human error—seams improperly sealed, holes poked by heavy equipment. Leachate collection pipes become clogged with silt or mud, or are blocked by the growth of microorganisms or the precipitation of minerals. Weakened by chemical attack, pipes are crushed by garbage.

Landfill covers fail, too: freezing and thawing cycles erode them; plant roots try to penetrate them. Woodchucks, mice, moles, voles, snakes, tortoises, ants, and bees innocently attack the cover from above, while buried tires, which have a habit of rising from the dead, threaten from below. If the cover becomes exposed, sunlight dries out clay layers and destroys plastic membranes; subsidence, caused by settling of the waste or organic decay, can result in cracks, too. So fragile are these systems, say landfill opponents, that state-of-the-art landfills merely delay, rather than eliminate, massive pollution to groundwater. The classical dump, at least when it comes to protecting groundwater, may even be preferable to modern landfills, where monitoring stations are spaced too far apart to detect fingerlike plumes of leachate springing from tiny holes. The unlined dumps leak evenly, said G. Fred Lee, and you know pretty quickly when your off-site groundwater becomes tainted by garbage.

The EPA requires landfill owners to monitor their sites for thirty years postclosure to control leachate and methane buildup, which causes fires and explosions. After this period, there is no funding to monitor water or air, to maintain landfill covers, or to remediate any eventual pollution. Over time, landfills pose more of a threat to the environment, not less.

Our hands were warmed up now, and I asked Alderson what effect the dump had on the nearby plants and animals, most of which had evolved in the absence of leachate. "Nitrogen and phosphorous are essential nutrients, though they can also be contaminants," he said. "But the net effect seems to be beneficial. Our spartina's growth exceeds or is equal to growth in a natural marsh."

I couldn't hide my surprise: after all, raw sewage flushed into creeks and channels upstream, and the tides swept umpteen kinds of poison from the landfill into the water twice a day. And it wasn't as if the Arthur Kill, into which Main and Fresh Kills Creeks flowed, was sluicing the place clean on every flood tide, either: there were twenty-five hundred acres of old and unlined landfills in the Meadowlands, just up the harbor from the Arthur Kill, out of which flowed a little more than a billion gallons of leachate a year.

But Alderson was sanguine about all this. He judged the marsh's health by its productivity, which he measured by colonization of benthic invertebrates, like ribbed mussels, by colonial wading birds, and by fish. He was happy to report the presence of baby bluefish, raptors, and egrets. Such resilience astounded me, though I'd later learn that cadmium and other persistent pollutants from the water were showing up in bird eggs and chicks, possibly jeopardizing their long-term survival.

Today, though, Alderson was showing off a restoration triumph. We continued paddling south, toward an empty barge and a mesh fence that stretched across the channel to a bulkhead. The creek was wider here and the current a little stronger. Alderson had warned me on the phone that debris barriers and booms would separate me from my dream of paddling around mounds of trash. But suddenly, he became animated. "Hey, it's open," he shouted. If we kept to the eastern shore, we could just sneak through the opening in the barrier. "I've never seen this open before," he said, angling us toward the bulkhead and ducking as we glided under a slimy rope. Beyond the bulkhead was exactly what was supposed to be out here—a gently sloped, perfectly functioning tidal marshland.

"We codesigned a three-acre restoration, with a thousand linear feet of shoreline," he said. "The grasses act as erosion control and natural filters. They pull out excess nutrients, metals, pesticides, and herbicides. They also mitigate storms and provide plant and animal habitat." We couldn't see it from sea level, but the sanitation department had also created a surface-water capture basin in there, made of berm and rock. It acted as a secondary means of purifying runoff by capturing surface sediment before it hit the creek.

In Alderson's perfect world, all the waterways around the landfill, where ecologically appropriate, would be planted with salt marsh vegetation. "Landfill engineers didn't have this in their consciousness before," he said. "We brought them square into a new paradigm."

Going north took longer than going south. It wasn't the wind that made it difficult but the water: the tide was going out. The mudflats on either side of us, though I couldn't be sure, looked twice as broad as they had twenty minutes ago. I wondered how we'd get the boat all the way back to the Toyota. We paddled doggedly for ten minutes, until a loud and scolding "Yo!" drew our attention to the western shore.

A sanitation sedan was parked on the lowest contour of Section 3/4. "Shit," Alderson said.

"Paddle over here," shouted a figure in uniform. "What do you think you're doing?"

Alderson bellowed his name and Parks Department affiliation, then said that he'd been checking on his salt marsh. The san man wasn't impressed: "I don't care who you are, I'm giving you a summons."

"I can't get over there!" Alderson shouted. There was fifty feet of leachate-suffused mud between solid ground and the cleat on our bow. Even if we had clear sailing and weren't afraid of running out of tide to reach our takeout point, the idea of presenting ourselves to receive punishment beggared common sense.

"Where are you parked?"

"Travis."

"Get your boat over here! You are trespassing on—"

I didn't hear the rest. Alderson had morphed into an angry beast.

"I built that fucking salt marsh for you!" he shouted without a hint of levity. I could feel his anger coursing through the rotomolded plastic of the boat. "You're telling me I can't paddle in public waters from a wildlife refuge around a landfill owned by the city?" All that shouting interfered with his stroke. I sank a little lower in my seat but continued paddling away from the sanitation cop, in a gesture of moral support. "Talk to Phil Gleason!" he continued. "He'll tell you who I am. If you're gonna give me one summons, you gotta give me a hundred. Give me a hundred and fifty! I've been in your creek a hundred and fifty times!"

We were out of voice-shot now and the cop got into his car. Alderson instantly calmed down. "I'm sorry," he said to me. "I was reverting to the language of the landfill."

Chapter Three

Stalking the Active Face

I drove out to Pennsylvania on the first warm spring day of the year, full of zeal for my quarry and eager to leave the winter-weary city behind. Route 78, known as the garbage interstate, was abuzz with eighteen-wheelers hauling putrescible waste across New Jersey to one of Pennsylvania's fifty-one landfills. Pennsylvania imports ten million tons of waste per year from neighboring states, more than any other state in the union. The garbage traffic fills state coffers with surcharges (in 2002, fees for out-of-state trash added up to $40 million), but it also brings danger. During one eight-day crackdown in the spring of 2001, inspectors found 849 trucks with violations so serious they were ordered off the road. Altogether, 86 percent of the more than 40,000 trucks inspected had safety and environment-related violations, such as leaks or improperly covered loads.

The landscape of rural Pennsylvania bespoke the geography of iron ore—gouged hills and twisting hollows of second-growth hardwoods, the towns depressed and faded. My destination lay a few easy miles north of Route 78, on the bucolically named Applebutter Road. I had figured Applebutter for a new road that cut through a former apple orchard. Instead, it was an old road that still had its orchards, and its eighteenth-century stone farmhouses, and its hillsides of goats and sheep. But across the road from the sheep was an enormously ugly power plant, and all around the plant were the rusty dregs of Bethlehem Steel—railcars and tracks, blackened hulks of hangars and factories. The Bethlehem property was the largest brownfield, or industry-polluted site, in the state.

My plan was to follow my kitchen trash to its final resting place, at the landfill in Bethlehem. The dump was owned by IESI, the same company that owned my transfer station down on Court Street, which I thought would make things easy, in terms of getting access. But I was wrong. Nothing in garbage was easy.

The landfill's manager, Sam Donato, didn't return my calls, so I phoned his boss, in Texas. Mickey Flood was IESI's founder and CEO, and he'd been friendly when we first spoke, months earlier. He'd griped to me about New York's rate cap on commercial carting: he said he couldn't make any money hauling restaurant waste, which was wet and heavy, if he could charge only $12.20 a yard for it. "It costs me fourteen dollars a yard!" Flood said. "But I know someone's picking it up, and they're either philanthropists or they're cheating." I had sympathized with Flood's predicament, and he had cleared the way for me to visit the Court Street station.

This time, Flood wasn't so helpful. In fact, he didn't return a single one of my calls, so I continued to hound Donato until he accidentally picked up his phone one day and reluctantly agreed to a tour.

Donato hadn't been warm on the phone, and he wasn't warm in person, either. He was a medium-size man with puffy cheeks, dressed in a white oxford-cloth shirt with the IESI pine tree logo. We met in the landfill office, a prefab building with vinyl siding. I sensed that Donato was rushed, so I tried to minimize my questions—a little landfill history, its capacity, and so on. When I had signed the register at the front desk I noticed that a groundwater-testing company had recently made several visits. I asked Donato about his leachate discharge, and he said, "I can't answer that, technically." The cell phone in his pocket kept buzzing, and he kept answering, chipping away at our time together. I cut to the chase: "Maybe we should just go up and see the landfill and I can ask you questions while we drive."

"I'll show you the recycling area, but you can't see the landfill," he said, leading me outside.

"But you said you'd show me the landfill," I said.

"I'm a little busy today."

"Could someone else show it to me? I just want a quick peek at the active face."

"No, no one else can show you. There's nothing to see—just look up there." He pointed uphill to Area 3D, a five-acre section way up at the top of an ocher-colored mound. A tractor-trailer was grinding its way through the lower gears, up one switchback and then another. I saw the truck's profile against the blue sky, then a puff of black exhaust escaped from its tiny stack and lingered over the cab like a cartoon cloud.

And that's all I saw. I assumed an excavator dug out the truck or maybe a hydraulic piston pushed a plate out and the compacted garbage spilled onto the ground. It didn't really matter what it looked like: anyone who reads magazines or watches TV can picture a vast amount of garbage, can understand the enormity of First World wastefulness. Or so I told myself. But I really did want to experience the whole thing. After all, it had been my garbage.

"When we spoke on the phone you said I could see the active face. I drove three hours this morning to get here."

"I'm sorry about that, but no." Donato didn't look the least bit sorry. In fact, I think he was smiling to himself. "I'll show you the recycling." He started walking toward a Dumpster behind the weigh station, but I stood my ground in the parking lot.

"I don't care about the recycling," I said. "It's not my plastic or metal or glass. I want to see where you're dumping trash, my trash."

"It's just too dangerous," Donato said.

"I won't get out. I'll ride up with a truck driver if you don't have time."

"No. There are liability issues. Those are private trucks."

"But you're vertically integrated! Those are your trucks. And I've already talked to Mr. Flood. He let me into the transfer station; he knows what I'm doing."

"There are no IESI trucks from New York today."

"Then I could ride with another driver."

"No."

I was running out of steam. "Okay, well, tell me this: is it a big hole up there or are you building layers?"

"Layers. It's flat. They try to work in small sections, maybe a hundred feet by fifty. Then they cover it. Or they put a tarp over it. We can do that."

"How high will it be when you're done?"

"We have a plan. There's a final heighth [*sic*]."

"All right."

"Thanks for coming out," he said jovially.

"Yeah."

I got into my car and started writing notes, hoping to make Donato, who still stood in the parking lot, nervous. I used the Bethlehem landfill pen I'd picked up in his conference room. When Donato went back inside, I went driving around. I talked to his neighbors about the smell of the place, the trucks on the road, the gulls kettling over the trash mountain, which was really—from a distance—a simple-looking thing, Bactrian and brown. I viewed the hump from the south and the west and points in between, trespassing on dirt roads and around the backyards of cabins. The trucks were dumping toward the north and the east, but there were no roads near those heavily wooded slopes. I wanted to find a way to get into the landfill on foot, but a tall chain-link fence topped by three strands of barbed wire surrounded it, and the woods were heavily posted.

After my sixth pass down Applebutter Road, I noticed a creek flowing out of the landfill property and under the chain-link fence.

I parked near a cabin, hoping no one would call the police about my out-of-state plates, and dove into the woods. Darting from tree to tree, I made my way to the creek, crossed it on a rotting log, wriggled under a gap in the fence, and headed north, toward the rounded mounds of trash.

My dull canvas coat was good cover in the early-spring woods, and it was also good protection against wild raspberry and rose thorns. The landfill, it turned out, was like Sleeping Beauty's castle, protected on its lower slopes by a thick overgrowth of spiky brambles. A cottontail rabbit and a white-tailed deer bounded away from me. I found a length of deer vertebrae in the leaf litter. With a start, I realized I was having a wilderness experience right at the edge of a landfill. There was actually very little trash in here—and no human footprints, beer bottles, or condoms, baseline evidence of recreation—which made it easy to imagine I was part of an elite minority, one of the few civilians to hike these woods in modern times.

Closer to the forest's edge I crept. When I hit the landfill proper—the place where the ground started to rise and no trees grew—I made a quick reconnaissance. No one was in sight. Winter-matted brown grasses covered the slope. I ran up a small foothill, dodging an aloe plant and brushing past goldenrod and Queen Anne's lace. At the top, I peeked over the edge, prairie dog style. Damn! A white pickup was parked on a switchback a quarter mile up. I ducked down and retreated to the woods, then plotted an easterly course and reemerged onto the landfill. Again, I crept up a foothill. Another truck! Were they watching me? I regretted not bringing my binoculars. As I tiptoed back into the woods, I heard a tractor-trailer coming down the switchback. Donato had said that seventy trucks entered the landfill each day, dumping 750 tons, but only fifteen to twenty of them were eighteen-wheelers. By modern-day standards, the Bethlehem landfill was petite. Over in Morrisville, Pennsylvania, Waste Management was doing more than thirteen times that amount.

I waited for the pickups to leave. A turkey vulture circled incessantly just fifty feet overhead. "Go away," I whispered, beginning to feel nervous. I imagined Donato appearing from the brambles and grabbing me by the ear. Why did he care so much about keeping me out? I knew now that I'd never make it up to the active face on my own: there were five switchbacks to clear in half a mile of open terrain, the two lurking pickups, and the rest of those tractor-trailers rumbling up and down. Why was this landfill being guarded like a strategic target? Many small-town dumps were still open to the drive-in public seven days a week. I had recently seen a movie in which Richard Gere careens into a Westchester County dump in the dead of night to tip a body wrapped in a carpet. (Naturally, the body was discovered the next morning: all garbage workers have finely honed search images for bodies rolled up in carpets.) But that was Hollywood. In the real world, garbage had become increasingly privatized and removed from the public eye. Where the town dump had once been open, an area of social congress, landfills were now fenced and gated, so that fees could be charged and large equipment could operate freely. Gleaners and scavengers, who'd once performed a useful public function, were no longer welcome. The garbage was still garbage, but it was hidden—except from those unlucky enough to live near the landfill's edge or downwind from its active face. People preferred their waste out of sight, but that distance also cultivated secrecy.

When I got home I phoned Al Wurth, the political scientist from Lehigh College, which was just a few miles from the Bethlehem landfill. Wurth had a lot to say about my naïveté. "They may arrange a tour for a class of third-graders, but they're not going to let a writer in," he said. I asked if there was a citizens group that had been fighting the landfill; he said that when the township signed its host agreement with IESI, such activism had been forbidden.

"Geez," I said, frustrated. "I mean, what have they got to hide?"

"Look," Wurth said significantly. "This isn't goods they're transferring from place to place. This is bads."

I knew the Bethlehem landfill wasn't permitted to accept hazardous waste. So maybe the issue wasn't what they handled so much as who was doing the handling. According to a June 1986 federal report entitled "Organized Crime's Involvement in the Waste Hauling Industry," "there is a substantial body of evidence that organized crime controls much of the solid waste disposal industry in New York State and elsewhere." Of the $1.5 billion taken in each year by New York carters, estimated Peter Reuter, author of the 1987 report *Racketeering in Legitimate Industries,* commissioned by the Rand Corporation, "about 35 percent was excess sucked out of the customers by illegal activities."

Still, Benjamin Miller's *Fat of the Land,* a rich history of garbage in New York City, runs for nearly four hundred pages without mention of organized crime. William Rathje and Cullen Murphy, in *Rubbish!,* also ignore the topic, as did Robin Nagle when she taught a class called "Garbage in Gotham" to her anthropology students at New York University. "I didn't want to bog down in it," she told me. "I thought it might be a distraction."

I felt the same way. What, after all, did the Mafia have to do with my trash, my san men, and the goings-on at the Brooklyn South 6 garage? My understanding was that the mob stuck its pinky-ringed fingers into commercial hauling but left residential refuse—the province of uniformed, unionized civil servants—alone. What I hadn't fully realized is that even before Fresh Kills had closed, some of the city's recyclables had ended up in the same place as commercial waste: in transfer stations run by a local carting company whose principals were later convicted of organized-crime activities.

In the early nineties, Manhattan district attorney Robert Morgenthau, who had been investigating organized crime in the garment industry, began poking around the private carting business. His investigations coincided with the rocky entry into the New York market of Browning-Ferris Industries (BFI), whose main business strategy was to undercut the going rate. Angered by this tactic, mobsters began harassing BFI clients, sending their own trucks to pick up the garbage from office buildings and restaurants before BFI could get there, then returning the clients' garbage. The mess would bring hefty city fines. Mob drivers yelled insults at BFI drivers. Goons stole and vandalized BFI equipment. Once, a mob-operated truck tried to run a BFI truck off the road. Two-thirds of the customers that signed up with BFI backed out of their deals after visits from representatives of their former haulers. One morning a BFI supervisor got a call from his wife reporting that someone had dumped the head of a large German shepherd on the lawn near their mailbox. Taped in the dog's mouth was a note that read, "Welcome to New York."

Enough was enough, and Browning-Ferris agreed to work undercover for Morgenthau as it trolled for new clients. Using bugs and wiretaps, the DA had by June of 1995 collected enough evidence to hand down a 114-count indictment against twenty-three carting companies, seventeen individuals, and four trade associations for antitrust violations, enterprise corruption, grand larceny, arson, assault (including a nearly fatal beating of a driver), and criminal conspiracy under the Organized Crime Control Act. The indictments also accused the companies of improperly disposing of contaminated waste.

While the indictments wound their way through the courts, Mayor Rudy Giuliani, a former federal prosecutor who had campaigned on a vow to run the mob out of town, formed a new city agency called the Trade Waste Commission. Its mission was to overhaul the commercial carting system and squeeze out mob-connected local companies. The Trade Waste Commission told private businesses they had the right to freely choose a carter and to cancel their contracts with thirty days' notice. It required carters to be licensed by the commission and forbade them from charging more than the set rate.

The local carting industry immediately challenged the legality of the commission. Hearing the complaint in the US Court of Appeals for the Second Circuit, Judge Richard J. Cardamone was moved to reference Stephen W. Hawking's *A Brief History of Time:*

> *Like those dense stars found in the firmament, the cartel cannot be seen and its existence can only be shown by its*

> *effect on the conduct of those falling within its ambit. Because of its strong gravitational field, no light escapes very far from a "black hole" before it is dragged back. . . . [T]he record before us reveals that from the cartel's domination of the commercial waste industry, no carter escapes.*

The complaint was overturned.

Giuliani assigned thirty police detectives to investigate corruption, inspectors to root out suspected overcharging, and auditors to examine the financial dealings of the carting companies. During its first year, the Trade Waste Commission oversaw the shutdown or sale of nearly two hundred garbage-hauling companies and denied licenses to many others. The commission claimed to have saved city businesses more than $500 million in inflated trash bills a year. According to Detective Rick Cowan, in *Takedown: The Fall of the Last Mafia Empire,* about his undercover role in the BFI sting, the bill for the Empire Blue Cross / Blue Shield building on Third Avenue fell from $650,000 a year to less than $80,000; the World Trade Center's bill dropped from $3 million to $600,000; the bill for an office building on Water Street plummeted from $1.2 million a year to $150,000. Eventually, all of the individuals, corporate entities, and trade associations indicted by Morgenthau either pleaded guilty or were convicted by trial jury. The defendants paid a total of $43 million in fines.

The cleanup of the private carting companies created a vacuum quickly filled by the large national waste-hauling companies (including IESI, Waste Management, and Allied, which soon acquired BFI), many of which already operated local transfer stations, area landfills, and materials recovery facilities (known as MRFs, pronounced "murfs"), and at least one of which, Waste Management, had a history of price fixing, bid rigging, insider trading, fraud, and environmental violations. By the late nineties, though, the companies started complaining of big losses: they wanted to charge more for their services. Giuliani refused to raise carting rates significantly, but toward the end of his term it became obvious that for the MRFs, at least, to operate efficiently, they had to raise rates high enough to make the capital improvements that would let them capture and process more recyclables, to say nothing of bringing their facilities up to city and state standards. (The MRFs were notoriously dirty, dangerous places. The *New York Daily News* reported three accidental deaths in 1996 at Waste Management's transfer and recycling facility in the Williamsburg neighborhood of Brooklyn. In 1999, a severed human head—badly damaged—showed up on a conveyor belt in the same sorting center, though it probably had nothing to do with a workplace accident.) Among the reasons Mayor Bloomberg cited for suspending plastic and glass recycling in 2002 was to avoid paying the huge rate increases that Waste Management and Allied insisted upon. The city had been paying roughly $58 a ton to drop recyclables at their MRFs; the new rate would be $120.

"There was a lot of consolidation after Giuliani ran the mob out," Robert Lange, of DSNY's Bureau of Waste Prevention, Reuse and Recycling, told me. "And lots of buyouts, and now we have a monopolistic system and skyrocketing costs to export garbage." Between 1996 and 2002, the Department of Sanitation budget nearly doubled, from $631 million to about $1 billion. Meanwhile, back in the realm of private carting, the national companies' prices inched up and up until they reached historic Mafia levels. Now, according to private carter Sal Benedetto, "the only difference between the majors"—the nationals—"and the boys"—the mob—"is that the majors don't actually kill you."

I called Flood three more times and left messages with Suzy, his extrafriendly secretary. Eventually she suggested I call Ed Apuzzi, my old transfer station friend. I'd already left him several messages, but just for sport, and using Suzy's name, I left one more. While waiting for my phone to ring, I biked down to the transfer station to take a look at the drivers waiting in the tractor-trailer queue. Might one of them let me ride with him into the landfill? I spent some minutes shopping for a trustworthy face and a clean-looking cab, then performed a reality check and went back home.

And then one day, Apuzzi called. My heart skipped a beat. I told him the troubles I'd been having, and he told me Bethlehem had just gotten a permit to expand operations. Maybe that explained why Donato hadn't wanted visitors earlier.

"So it's okay for me to see it now?" I asked.

"Oh, no," Apuzzi said. "For insurance purposes, you can't walk around on the landfill. It's just not our policy."

"I don't want to walk around on the landfill," I said. "I just want to see it, from the front seat of a truck or something." I told him I'd stay put in the vehicle and sign any waivers he wanted.

He paused. "Let me check with some lawyers. I'll get back to you." I said thanks and hung up, knowing I'd never speak to him again.

Why was it so hard to look at garbage? To me, the secrecy of waste managers—which was surely based on an aversion to accountability—was only feeding the culture of shame that had come to surround an ordinary fact of life: throwing things away. Sure, the volume was shameful (especially the volume of stuff that could have been reused or should never have been acquired in the first place), but the volume wasn't Ed Apuzzi's or Mickey Flood's, it was ours. And yes, garbage wasn't a pretty sight, but so what? The sewage treatment plant gave tours, after all. Sanitation workers wanted civilians to own up to their mess, to take responsibility for their discards, to show some respect for those who dealt with their waste on a daily basis. I was in complete concordance. But I found that from the moment my trash left my house and entered the public domain—where no higher authority than the US Supreme Court had determined it was open for general inspection—it became terra incognita, forbidden fruit, a mystery that I lacked the talent or credentials to solve. I was left to imagine the worst: that the Bethlehem landfill was casually strewing the waterways and roadways with litter, or commingling municipal solid waste with leaking barrels of hazardous chemicals, or disposing of second-rate starlets. (Bodies did show up in landfills from time to time, and they weren't necessarily mob whack jobs. Seeking warmth and safety, the homeless sometimes slept in Dumpsters, and they sometimes got crushed by garbage trucks.)

By now, I was asking everyone I knew in the waste business if they could get me into a landfill: the folks at the company that handled the city's metal recyling; my contacts at DSNY; the manager of my former recycling facility; my Gowanus paddling partner, who knew someone whose family ran tugs in the harbor. In my desperate state, even that seemed promising. But nothing came of these probes.

I continued to cogitate. The popular conception of "bads" aside, I knew that the ordinary discards of residential life were hardly inert, that burying things under several feet of dirt didn't bring their influence on the environment to a screeching halt. When organic matter decomposes, it creates methane and carbon dioxide, both greenhouse gases. As it filters up through layers of buried garbage, methane can pick up carcinogens like acetone, benzene and ethyl benzene, xylenes, trichloroethylene, and vinyl chloride. These compounds are borne on the breeze into nearby homes and offices.

Just as jurisdiction over trash moved from the public realm to the private, its environmental impact moved from the local—right here in my kitchen—to the general. On its own, my trash might be harmless to me, but combined with the output of several million others, it could be lethal to many. A 1998 New York State Department of Health study found that escaping landfill gases contributed to an estimated fourfold increase in bladder cancer and leukemia rates in women who lived within 250 feet of thirty-eight upstate landfills. Researchers at Imperial College in London reported in 2001 that children of parents living near landfills in England tended to have a higher rate of birth defects than the general population. (Neither study could prove a direct cause-and-effect relationship between exposure and disease or defects, though both called for further study.)

In a few months I'd visit a medium-size landfill in Dixon, California, and study a schematic of its man-made stratigraphy. Directly underneath the "operations layer"—a foot-deep blend of soil and dried sewage sludge that underlay the trash—was six inches of gravel, followed by a layer of geotextile fabric, and then another bed of gravel (to collect and remove leachate). Next came a high-density polyethylene geomembrane, a geosynthetic clay liner backed with another layer of high-density polyethylene geomembrane (forty millimeters thick—the height of forty stacked dimes), a twelve-inch layer of compacted clay, six inches of compacted soil, a six-inch capillary break, and then, at the superbottom, a compacted subgrade of sand. The liners had tough-sounding trade names like Bentomat, Bentofix, Claymax, Geolock, and Geoflex. I'd asked Greg Pryor, who was giving me the blackboard tour, if the whole shebang keep the groundwater and soil safe.

"We get 99.3 percent protection," he said with pride.

"Do you aim for 100 percent?" I asked.

"You've got to look at cost. We pay $210,000 per acre to build landfills now." (Pryor wasn't counting all the expenses: according to Waste Management's Judy Archibald, permitting and construction costs for landfills, in 2002, ran $500,000 an acre.) In short, he didn't think the ratepayers would make 100 percent containment possible.

Groundwater contamination is a serious issue, but landfills are a nuisance in a myriad of other ways. Trucks drop litter and drip leachate on feeder roads, tugs pushing garbage barges pollute waterways, and clouds of scavenging birds darken the dump's dust-filled skies. Some landfill messes are relatively benign, some aren't: dumps seem to attract environmental lawsuits like flies. As of June 2003, 413 of the 1,571 sites on the EPA's National Priorities List, representing the worst of the worst Superfund sites (which by definition contain hazardous waste) were landfills, a ratio of just over one in four.

IESI Bethlehem wasn't on the Superfund list, nor had it received any notices of significant violations of the Clean Air Act, Clean Water Act, or Resource Conservation and Recovery Act within the last two years. But then I looked back a little further. I learned that IESI had owned the Bethlehem landfill only since 1999, when it had purchased it from Waste Management for $65 million and an agreement to pay Lower Saucon Township (which included the city of Bethlehem) a host fee of $460,000 a year. A hot potato, the Bethlehem landfill had changed hands, just before IESI had bought it, three times in less than a year. There had been questions raised about groundwater contamination before IESI's watch, and complaints about overweight, leaky, and generally unclean trucks. Within months of purchasing the landfill, IESI had an even nastier surprise. Workers excavating an old section of the property discovered 280 fifty-five-gallon drums and 55 five-gallon cans filled with lead, PCBs, creosol, and the solvents trichloroethylene and tetrachloroethylene. The drums, which had probably been buried in the fifties and sixties, were punctured and leaking. IESI paid for the cleanup: the waste was trucked to New York for incineration.

Two years passed, and IESI petitioned Lower Saucon Township for permission to expand its daily take from 750 tons to a maximum of 1,800. To sweeten the deal, the company offered two thousand-dollar scholarships to local high school students. Asked at a town hearing what would prevent someone from trying to hide hazardous waste in the bottom of a garbage truck, IESI's consulting engineer answered, "Probably your conscience."

Originally, the company had proposed increasing its daily maximum volume to two thousand tons. When Lower Saucon countered by advertising an ordinance that would have reduced the landfill's Saturday hours and limited its daily maximum to one thousand tons, IESI threatened to sue the township. Apparently, their old host agreement had stipulated that the township could not oppose future increases. Wary of environmental impacts, the town next hired independent consultants to examine the landfill and asked IESI to conduct further studies on geology and search for contaminants. Pennsylvania's Department of Environmental Protection ruled against further study and finally, in the spring of 2003, granted the expansion.

I still wasn't getting anywhere running my own garbage to ground, so I took a little break and tracked the trash of neighboring Manhattan. Most of the five boroughs' waste traveled to landfills far and wide, but about 60 percent of the stuff collected at Manhattan curbsides wasn't buried anywhere. It was burned in the American Ref-Fuel waste-to-energy plant in Newark, New Jersey. One morning I drove out to the exurban wilds of the Meadowlands, which in pre–Fresh Kills days had been the largest garbage dump in the world, to see what this modern process looked like. As opposed to old-fashioned incinerators, which burn waste with comparatively few environmental controls, waste-to-energy (WTE) plants are technologically evolved contraptions that convert trash into energy while meeting state and federal air-quality guidelines. Across the country, there are now eighty-nine WTE facilities operating in twenty-seven states. Collectively, they burn 13 percent of the nation's garbage.

Surrounded by a tangle of potholed highway ramps, littered underpasses, brick warehouses with shot-out windows, and the phragmites-covered landfills of yesteryear, American Ref-Fuel looked sleek and modern under its skin of rose- and blue-colored panels. The parking lot smelled a little like a pigsty, thanks to the parade of garbage trucks, but the executive offices, where I met plant manager Jim White, smelled almost normal. White sat me down in a conference room filled with leather chairs and a long shiny table, then showed me a corporate video that explained, in language a ten-year-old could understand, the virtues of modern incineration. He clicked off the tape before the credits rolled and led me, in brisk silence, through a heavy door into a windowless concrete passageway. We clomped down a steel staircase to the grandly named tipping hall, where packer trucks pulled in and disgorged their loads onto a concrete pad. Eagle-eyed workers plucked large chunks of iron and steel, plus furniture, from the aromatic jumble. Over the course of a year, they'd salvage some forty million pounds of metal. (At an American Ref-Fuel sister plant nestled among scrub pines on Cape Cod—a facility that served a sixty-five-square-mile area and took the place of forty town dumps—big metal was pulled out, and then the garbage was shredded and run past magnets, which stripped out smaller chunks of metal, before it was subjected to flame.)

I watched a bulldozer push Manhattan's trash, mingled with that of Essex County, New Jersey, into a concrete bunker three hundred feet long, ninety-five feet high, and seventy feet deep. When full, the bunker held thirteen thousand tons of trash, nearly the amount generated in a single day by New York City residents. (Commercial waste, remember, accounted for an additional thirteen thousand daily tons.) The stockpile would feed the fires on Sundays, the only day of the week that fresh garbage wasn't delivered.

White and I climbed back upstairs to take a peek at the control room, where a young man clicked through a series of computer screens. His business was numbers, the quantification, in grams and degrees and tons, of American Ref-Fuel's basic elements: air, water, fire, and garbage. Three color monitors featured continuous video feed from the glowing belly of the trash-burning beast. Mesmerized by the computer images, I took a photo, and then White brought me into a glass-walled control booth where a grapple operator named Romeo made tiny joystick movements that resulted in the fluffing of the bunker's malodorous contents. "Believe it or not, there's an art to this," White said. "We're making a unique combustion product." Down below, the grapple steadily fed product in eight-ton bites, into a one-hundred-foot-long boiler on a 35-degree slant. We clomped back downstairs to its base. When I leaned toward a small glass window to see, face-to-face, what a 3,000-degree trash fire looked like, White barked, "Don't touch!" The window was hot. The walls of the boiler were lined with water-filled tubes, he said, and while the garbage burned on roller grates and shrunk to a quarter of its original weight, it generated steam that drove two 50,000-horsepower turbines to create electricity. "We're producing sixty-seven megawatts an hour right now," White said, enough to power fifty thousand homes.

It sure sounded like a good idea: burn waste, get energy. "We've come a long way since the incinerators of the seventies," White said. He described a complicated pollution-control recipe that involved scrubbers, electrostatic precipitators (to charge particles so they could be collected), flue-gas cleaning, combustion controls that minimized carbon monoxide, and injections of carbon (to absorb mercury), lime (to control sulfur dioxide and hydrochloric acid), and ammonia (to control nitrogen oxides).

But somehow, it wasn't enough. Burning a mixed stream of natural and synthetic materials creates newfangled compounds that release dangerous gases. Plastic, for example, releases hydrochloric acid (which rapidly degrades the incinerator and contributes to acid rain), chlorine (which is then available to form dioxins), and toxic metals that have been added to the plastics to give them color, stiffness, and other desirable characteristics. Scrubbers and screens catch much of this stuff, but even minute quantities, once airborne, are considered by scientists to be extremely dangerous. Improved technology and higher air-quality standards have taken some metals—like chromium, copper, manganese, and vanadium—out of the smokestack only to concentrate them in bottom ash, which falls through the grate on the boiler's floor. Though acknowledged to be toxic, the combined bottom and fly ash (which consists of fine particles removed from the flue gas by all those screens and filters) is dumped not in hazardous-waste landfills but, thanks to an EPA exemption, in landfills designed for household waste.

White told me the metals in his incinerator ash were "locked up" by these physical processes and therefore inert. Opponents of WTE believe this condition is only temporary and that toxins will eventually leach out, especially if the ash is combined with other materials and used in construction projects. This is just one of the so-called beneficial uses for ash that incinerator operators are pursuing. Remember, they pay to tip their waste product in landfills, too. Incineration reduces the weight of household trash by 75 percent, but that still leaves 3,250 tons for every bunker-load burned—hardly the discrete box of cremains that promoters of WTE would have you imagine such plants produce. The potential health and environmental impacts of beneficial reuse are unknown. New Jersey forbids landfills to use incinerator ash as daily cover, but other states aren't so fussy.

The Newark plant had been running since the early nineties. Starting in the eighties, New York City had tried to build its own waste-to-energy plant, at the Brooklyn Navy Yard, in Williamsburg, but community opposition defeated the proposal in 1996. More recently, Steven Cohen, of Columbia University's Earth Institute, proposed in a *New York Times* op-ed a regionwide solution for the city's solid-waste problems: ship city trash upriver to "depressed localities that dot the Hudson's shores," then burn it in waste-to-energy plants. Predictably, the idea sparked protest from those who lived in those depressed localities, as well as from recycling advocates capable of sniffing prejudice from three thousand miles away. "[T]he construction of waste incinerators in 'depressed' or 'industrial' locales has historically been associated with environmental racism," responded the Ecology Center, of Berkeley. "Poor people of color are most likely to work and live near these areas, making them the primary victims of incinerator pollution."

One afternoon I went up to Columbia University's Earth Engineering Center to speak with Nick Themelis, who had been studying waste for five years and was a colleague of Steven Cohen. Themelis had closely cut white hair and wore a green pullover with a zipper at the shoulder. "Engineers came up with the problem of waste, creating new things faster, so we thought we ought to solve it," he said. After looking at recycling and composting, "we realized that the benefits of waste-to-energy are crystal clear." I asked how New York City officials responded to the reams of data he'd collected from the world's most modern WTE plants. Themelis shrugged: he was fed up with their lack of interest. "I don't care about New York," he said. "If I can't do it here, I'll do it somewhere else." He had plenty of customers, in other nations, for his incinerator expertise.

While engineers and chemists, community activists and environmentalists debate, one thing is certain: incineration, perhaps even more than landfilling, competes with attempts to reduce our nation's enormous volume of waste. According to the Institute for

Local Self-Reliance, most incinerators require "put-or-pay" contracts stipulating that local governments deliver a guaranteed tonnage of material to the incinerator or pay a penalty. If you have to pay anyway, why would you bother to reduce, reuse, and recycle?

Incinerator proponents argue that waste-to-energy can work in concert with recycling (in fact, says Alan Eschenroeder, of Harvard University's School of Public Health, communities with WTE plants recycle on average at a rate higher than communities that landfill), and that there is enough garbage for everyone. The anti-incinerator crowd amps the fear factor: the unknown harms of persistent and bioaccumulative toxics, even if they do occur in quantities far lower than those of a decade past. (According to Themelis, emissions of dioxin, mercury, cadmium, lead, hydrochloric acid, sulfur dioxide, and particulate matters from WTE facilities dropped between 86.7 and 99.7 percent between 1990 and 2000.) But maybe the numbers don't matter. Over the past decade, community opposition and tighter federal and state regulations have made siting and building waste-to-energy plants extremely expensive—about $120 million for a facility that processes just a thousand tons of trash per day. In fact, the US hasn't seen a new one since 1996. (Other nations are more gung ho. Since 1996, 165 WTE plants have either been built or are under construction abroad.)

Arranging to witness the bonfire of Manhattan's detritus had been refreshingly simple, but my quest to visit the IESI landfill in Pennsylvania was looking more and more quixotic. I decided to further broaden my parameters. I no longer cared about seeing my own trash interred in the ground: I'd be happy to look at Staten Island's or even New Jersey's. Through an acquaintance of an acquaintance, I got hooked up with Andy Reichel, an environmental consultant who had plans to visit the Monmouth County, New Jersey, landfill. He invited me to tag along.

A few weeks before the summer solstice, I sped past Fresh Kills, crossed into the Garden State, and wound my way east and south along the Jersey Shore. When I met Reichel in the parking lot, he warned me, "Just let me do the talking in there. Don't say anything about IESI not letting you into Bethlehem."

"Okay," I agreed hesitantly. "But what are you going to say when the superintendent asks why I'm here?"

"He won't," Reichel said firmly.

We met Chris Murray in a nondescript conference room adorned with landfill maps. If he was curious about my presence, here, he didn't let on. Monmouth County, we learned, had an unusual way of subduing municipal household waste: from the tipping floor, it went up a conveyor belt and into a hopper, which compacted the garbage into bales forty-five inches on a side. From behind a window thirty feet above the enclosed tipping floor, I watched as bulldozers shoved the county's household refuse into an area the size of three basketball courts. The garbage reached a height of twenty feet; a bulldozer zipped down a narrow canyon that split the pile through the middle.

The bales seemed made up mostly of plastic bags—black and white and yellow, all of them torn and dulled with grime. I picked out individual objects: a propane tank, a plastic flowerpot, paper plates, a bathroom scale, sheets of folded paper, a banana crate, a small toy shovel, a large plastic horse. I asked Murray if he'd seen anything truly interesting in there. He said, "Dead rats, dead raccoons, dead cats." Anything alive? I asked. "Rats, raccoons, cats," he said.

The architects of this building, which had cost $25 million to erect in 1996, had incorporated into the design a long L-shaped corridor with plate glass windows that overlooked the tipping floor, conveyor belt, balers, and recycling operation. The idea was that visitors would be allowed to see what happened to their trash: school groups would come through, senior citizens, city planners, book writers. The windows were big; little was hidden. The public looked and it learned. Today, though, there were no finger smudges on the window, and the hallway was dim. When the market for recyclables had dropped in the midnineties, heavy on supply but low on demand, National Ecology, the private company that held the recycling contract, had quit pulling plastic, paper, metal, and glass from the coffee grounds, pizza crusts, dust balls, and toys. Everything, now, went into the bales. The show had changed, and the school groups no longer came.

I watched forklifts position the compressed crazy-quilt cubes of waste on flatbed trucks—each cube weighed a ton and a half—then got into Chris Murray's SUV to follow them out into the light of day.

We drove on winding roads through the hundred-plus-acre property. Most of the landfill sections had been closed and planted long ago, so it was a bit like touring a country estate in a well-groomed park, but instead of waiting for the manor house to loom up from the monoculture of knee-high grass, I awaited a gash in the earth, the proverbial active face. Murray had the tour all planned: we circled around the back of the property and topped the highest mound, which had an elevation of 256 feet. A thousand-acre army base surrounded the landfill, and the view from up top was of unbroken forest, mostly oaks and pines. There wasn't a man-made object in sight.

Murray turned his Jeep around and we headed toward the main attraction. He parked on the muddy road, and we watched from a slight distance as forklifts snatched bales of garbage, one by one, from flatbeds. They turned and stacked them, three bales high, creating terraces of garbage on the edge of a flat plain. There was an air of solemnity to the garbage stacking, an aspect of restraint that befits a burial. But the men were, in fact, building even while they buried. I was reminded of Mayan ball fields and the ruins of other ancient cultures—the ochre color of the dirt, the escalating ramparts and terraces, the angularity of the place. I was looking at an ending, but I was thinking, This is how civilization began.

Archaeologists know that ancient people did the same things with refuse that we do today: they threw it in holes in the ground, they buried it, and then they found another hole. Sometimes they left detritus on the floors of their houses, then covered that layer with dirt. Over time the floors rose high enough that roofs and doorways had to be raised. The ancient Trojans, it's estimated, accumulated waste at the rate of 4.7 feet per century. In Don DeLillo's *Underworld,* a novel about a waste trader, a character named Jesse Detwiler teaches his UCLA students that garbage has its own momentum, that it has the power to shape people. Pushed to edges, garbage also pushed back. "People were compelled to develop an organized response," Detwiler says. "This meant they had to come up with a resourceful means of disposal and build a social structure to carry it out—workers, managers, haulers, scavengers." In Detwiler's view, garbage acted as an evolutionary force. "[I]t forced us to develop the logic and rigor that would lead to systematic investigations of reality, to science, art, music, mathematics."

Sitting inside the climate-controlled SUV, I sensed grandeur in the scene playing out before me. Here was Monmouth County's waste: this was Monmouth County's organized response. There was none of IESI's shame in the operation. It was just a place where men buried garbage.

Chapter Four

The Spectacle of Waste

Robin Nagle had invited me to join her undergraduate Urban Anthropology class at New York University, but I didn't take her up on the offer until one late-spring morning, when three employees from the Department of Sanitation had been invited to address her students. I was mostly keen on meeting Dennis Diggins, who happened to be the director of Fresh Kills.

Nagle began by walking the class through the officers' uniforms: the insignia, the hash marks on sleeves, the collar pins. For Nagle, uniforms were a text, or "identity markers." She wore her own set of these: a small barbell in the rim of her left ear, a tailored leather jacket, a shoulder tattoo. Nagle was tall and lean, a triathlete, and she favored long skirts, which lent her a sexy-librarian look. When her guests, describing their jobs to the class, used DSNY jargon, she interrupted peremptorily. "Explain RO," she'd command. (It meant rotating officer.) "Explain chart." (A san man's day off.)

Nagle, of course, knew exactly what these terms meant. She

had been working unofficially at the Manhattan 7 garage, in her Upper West Side neighborhood, for the last half a year. She woke before dawn and pulled on the green uniform, then spent the next several hours dragging garbage cans from curb to truck and tipping them inside. After "getting it up" with her two male partners—ten tons or so—she took the subway downtown to her paying job. The director of NYU's Draper Interdisciplinary Master's Program, Nagle investigated the social meanings of trash in her graduate-level "Garbage in Gotham" class and conducted ongoing ethnographic research on sanitation workers, about whom she planned to write a book. The following spring she'd take the citywide DSNY test, answer seventy-three out of seventy-five questions correctly, and go on to ace the physical.

Chief Diggins, as he was known, stood before the students with his green cap in his hands. There was a sadness to his ruddy, rounded face. His mustache was graying; his eyes were gentle. Diggins wasn't sure what Nagle expected of him, so he quietly offered an overview of Fresh Kills. "We finally got the landfill up to code, except for the liner, and then they closed it. It's kind of sad because we were a good neighbor—we controlled odors with gas and leachate systems. You don't smell anything now. It used to be awful." He didn't go into detail—and the students didn't press him—but for as long as state and federal environmental laws have existed, Fresh Kills had been violating them.

After Nagle's class I let a decent interval pass before phoning the chief at work and asking if I could come out to see him. I was apologetic, and he was accommodating. Nagle wanted a tour, too: she hadn't been to Fresh Kills since it had closed.

Keyed up, I arrived five minutes early for our 7:00 a.m. appointment. Nagle had been here for twenty minutes already. Her eagerness seemed in keeping with her competitive spirit. I noticed that we were dressed alike, in blue jeans, hiking boots, and white T-shirts. We both had black backpacks. We both had a four-year-old child home in bed.

"You bring something to eat?" she asked me.

"Peanut butter sandwich."

"Turkey," she said with an approving nod.

Chief Diggins had a large office inside a paneled trailer, one among a series of trailers stationed in a scruffy parking lot. From here, there was no view of trash, no hint of the marvels to come. A few trucks and cars rumbled around the potholed roads, driven mostly by private contractors building a new transfer station. DSNY had only about a hundred people on the payroll at Fresh Kills these days. (At its height of activity, the landfill employed more than five hundred, including metalworkers, chemists, blacksmiths, steamfitters, plumbers, riggers, welders, and machinists.) Things were so quiet now, the guard at the entrance gate was asleep in his tiny booth. "You gonna bang him?" Nagle asked Diggins, a gleam in her eye. She liked the word *bang:* it meant write someone up, get him in trouble. "Can't," said Diggins. "He's from a private company." But it was obvious the guard's days here—if not his hours—were numbered.

The phone rang and Diggins lifted the receiver an inch, but instead of saying hello he finished what he was saying to Nagle. I was struck by his self-assurance, a confidence that whoever was on the other end would not only wait but probably wouldn't mind listening to this conversation, either. Now Diggins said, "Diggins," into the phone, nodded a few times and said, "Uh-uh." He hung up, handled two calls on his walkie-talkie, then suddenly decided it was time we headed out.

The chief didn't start with the big stuff, the spectacle of waste. He drove Nagle and me through what would be, in a film studio, the back lot: miles of muddy roads connecting convocations of large machines. I saw front-end loaders with buckets big enough to hold a Humvee, dump trucks with tires higher than my head, bulldozers with studded iron treads. The lowlands, where the roads were flat, seemed elaborately unkempt. Chain-link fences were torn, vines crept over concrete barriers, weeds grew through cracks in asphalt. The wildness contrasted starkly with Monmouth County's trim little landfill, but I had to remember that burying twenty-six million pounds of municipal solid waste a day—or double that, in the days when Fresh Kills accepted commercial waste as well—was considerably more complicated than Monmouth's puny 3.7 million.

My first land-based peek at an actual section, closed for several years now, was anticlimactic: the graded road wrapped around an enormous shoulder of grass, and Queen Anne's lace, dandelions, black-eyed Susan, cinquefoil, yellow-blossomed mugwort, blue chicory, and pink multiflora roses covered its 145-foot-high plateau. There was a slight smell of swamp gas in the breeze, and meter-high candy-cane-shaped methane venting pipes sprouted from the ground here and there. But everything that was supposed to be important about this place—its Jovian mass, its horrifying contents!—suddenly was not. I got out of Diggins's truck on the section's plateau and, as the cops on TV say, took a moment. Not only couldn't I see the garbage piled under my feet, I had no other physical or emotional sense of it, either. I knew the place held 2.9 billion cubic yards of trash (about the volume of 1,160 Pyramids of Cheops), but someone had done a very, very good job of covering it all up.

What struck me, instead, was the incredible view. I could see for miles—over other hundred-foot-high mounds of grass, over creeks and marshes, over distant neighborhoods, and up over the treetops to the highlands of Staten Island. I felt a strong urge to ditch the tour and light out with a compass, to reconnoiter the entire 2,500 acres. This was, for 99.9 percent of the eight million people who lived nearby, completely new ground. In a vertical city that barely offered breathing room, this was a tremendous windfall of air, of unshadowed space, of landscape without edge, pavement, telephone poles, or people.

The tidal creeks were the best part of it. Fresh Kills, Main, and Richmond Creeks meandered around the bases of the great hills, forming a shaky letter *Y*. Rafts of bufflehead ducks and geese dotted the waters, which were buffered on either side by wide mudflats and ribbons of spartina grass. They gave the vistas perspective; they brought the eye in and out of the frame. Diggins said he was dazzled by it all, especially now that the place was closed and he wasn't constantly rushing from one emergency to the next. "When this place greens up, it's incredible," he said. "I always said that if you put a couple plastic cows up there"—he gestured across Main Creek toward a pastorally grassy slope—"no one would know the difference. You'd think you were in dairy country." In this bucolic paradigm, it was easy to imagine the two tall stacks of a methane flaring station as grain silos. (Well, almost. Imagining Fresh Kills as a dairy farm forced me to contemplate methane, a potent greenhouse gas, rising not only from the rotting garbage but also from the oral and anal sphincters of cattle, who, along with other domestic ruminants, annually produce about 19 percent of global anthropogenic methane emissions.)

Diggins seemed lost in thought. In the convivial silence, I scanned the middle distance and for the first time had an inkling of how a place that was universally reviled could be reborn—after a lot of fancy landscaping—as a place universally admired. A dump was a negative space, in every sense of the word. But a dump could be filled and capped, and something alive and new could be created in its place. Transformed into a positive space, the landfill had the potential to be, as city planners liked to say, an amenity.

We drove downhill and Diggins pointed out planted sections and sections recently closed but still sporting only a rough preliminary cover. All appeared calm on the surface, but I knew that underneath, the garbage roiled. Of all the landfills studied by the Garbage Project, Fresh Kills enjoyed the fastest rate of decomposition. Archaeology students digging in drier landfills have discovered forty-year-old hot dogs that look just like the ones currently sold in the Times Square subway station. Seventy-year-old newspapers can still be read. Cling Wrap still clings. Most landfills are more like mummifiers than composters, it turns out. Achieving a rich, moist brown humus in a sanitary landfill is nothing but a romantic fantasy! (As my garbage research proceeded, this fantasy would come to haunt me. Master composters dangled visions of "rich brown humus" before me; even Synagro, the company in the

Bronx that handles my neighborhood's processed sewage, tried to win me over by describing its product in those same exact terms.)

Of course, wet garbage starts decaying almost as soon as it hits the plastic bag in my kitchen—my nose told me that. But at almost every landfill in the country, this process grinds nearly to a halt once bagged garbage is compacted and buried. Below the top eight feet of a landfill, few organisms that require oxygen—which means precious few of the variety that most greedily chew up waste—can survive. (Where do those hungry microbes come from in the first place? Basically, everywhere: the air, the ground we walk on, the food we eat, the hands that tie up our garbage sacks.) Aerobic biodegradation works best when organic material is chopped up, kept moist and warm, and exposed to oxygen with regular turning. Any organic compound in this top layer of the landfill—or in the transfer station or the kitchen garbage pail—is fair game for digestion by bacteria, fungi, and insects, which use their enzymes to break the large organic compounds into fatty acids, water, and carbon dioxide. In this phase of biodegradation, the landfill temperature rises, and a weak acid forms within the water, dissolving some of the minerals.

When the aerobic microbes die off and oxygen is depleted, the anaerobic team takes over. The first wave of anaerobic bacteria produces enzymes called cellulases, which break organic material into smaller molecules, like sugars and amino or fatty acids. Next, acetogenic bacteria ferment those products into alcohols and organic acids—including acetic, lactic, and formic acids. The third and final wave of bacteria, the methanogens, converts acetic acid and methanol into underground plumes of methane, carbon dioxide, and water. If the gases escape collection hoses and rise through the layers of garbage, as they do in both old-style dumps and new, they feed potential fires and contribute to greenhouse warming.

Biodegradation works with microbes, but degradation—minus the *bio*—involves the breakdown of materials from chemical interactions. Think of oilcans rusting in the rain, or shampoo bottles exposed to sunlight becoming brittle with age and breaking into smaller parts. How long any decomposition process takes is an open question. Food and yard waste go faster: some items, like plastic and glass, never seem to go at all, especially in the absence of sunlight, wind, and water. The literature on landfill degradation is substantial, and it emphasizes the uncertainty of knowing when a closed dump will become stable. When the Austrian Environmental Protection Agency studied the issue, it apologized for considering a mere ten-thousand-year time frame, when a hundred thousand years is more likely.

The problem boils down to this: all burials are not alike. Temperature, pH of the fill and cover material, moisture level, and movement of fluids vary around the world, from day to day at any particular landfill, and even between different truckloads of waste disgorged on the same day. Our experience with six-pack rings and hybrid iced-tea cans is young, so we have little information about the degradation of newfangled materials either on their own or commingled with other materials, newfangled or not. Set and setting hold much sway: depending on its burial context, a Granny Smith apple can biodegrade completely in two weeks or last several thousand years.

Archaeologists, who've been systematically exhuming buried discards for well over a hundred years, have an acute interest in decomposition rates, which they call Natural Formation Processes, or NFPs. Among their surprising NFP discoveries is that ostrich plumes can remain intact more than three thousand years in a dry climate, and wooden carvings last nearly four hundred, if they're buried in very wet mud. It's understood that when moisture moves through a site, as in a bioreactor landfill, the rate of biodegradation increases. But when materials are saturated with water that doesn't move—in some swamps, in fill behind retaining walls in harbors, and even in outhouses, reports Garbage Project director William Rathje—textiles, leather, paper, and food have been found intact one hundred to one thousand years after burial.

Forensic anthropologists have studied how human bodies decay under very specific conditions: facing south in the sun at latitude 40 in the winter, for example, or facing north in the tropical shade, exposed to rain, sun, and wind. They know that "bog people" in British swamps can be perfectly preserved for 2,000 years, and that nomadic hunters frozen in Tyrolean glaciers can last for 5,300. In fact, we know far more about how our bodies decompose than about how the stuff that makes our lives possible, or pleasurable, decomposes. It's understandable: we've only been studying our waste self-consciously for less than a hundred years.

Why does an understanding of degradation rates matter? Besides our anthropological interest in discard dates (including post-mortem intervals), these rates let us predict when landfills will stabilize and, in a more perfect world, make informed decisions about whether we want to bury (or buy) something in the first place.

The majority of landfills excavated by the Garbage Project showed little evidence of biodegradation below the surface. The big exception, of course, was Fresh Kills. Why? Because it is very old, it isn't lined, it sits in a swamp through which channels and streams flow, and tides flush it twice a day. Even above the water level, garbage absorbs moisture like a sponge. In G. Fred Lee's microbe-chomping world of wet-tomb bioreacting landfills, Fresh Kills would be a stellar example, except for the fact that it doesn't have a floor. It is in no way a closed system.

Diggins continued narrating our tour. As the garbage at Fresh Kills shrank, he said, it generated methane, about twelve million cubic feet of it a day. Starting in 1996, the EPA required large landfills to install rudimentary gas collection systems, mostly to halt the frequent fires and explosions inspired by accumulating methane. For years, underground fires had burned out of control in the dumps of the New Jersey Meadowlands. At a 1986 concert at the Shoreline Amphitheater, built atop an old landfill just south of San Francisco, a Steve Winwood fan lit a cigarette and ignited a five-foot column of hair-singeing flame. Even when collection systems are installed, they don't always do the job. At one landfill refashioned into a municipal park in the Southeast, leaking methane migrated into a cavity beneath a concrete slab. When a soccer ball landed in this hole during a nighttime practice, the woman retrieving it flicked a lighter to get a better view inside. *Kaboom!* She survived the flash, which blew her (and presumably the ball) out of the hole, but her throat and lungs were badly burned. Of the states surveyed for *BioCycle*'s "State of Garbage in America" report in 2000, twenty-nine had at least one landfill recovering gas for energy, versus collection and flaring; fourteen had none; and seven offered no information.

Over the past couple years, Diggins said as he pulled over to let a dump truck pass, engineers at Fresh Kills had installed perforated pipes in the garbage mounds that sucked landfill gas into a boxy-looking plant near the West Shore Expressway. The gas was scrubbed of carbon dioxide, and the remaining methane supplied enough power to light fourteen thousand homes. At least that's what happened to half the collected gas: the other half, which was nonmethane organic compounds, was treated and then released. I asked Diggins what those distant flaring stations, with their fifty-foot-high stacks, were for. "We've got six flares," he answered. "They were used before all the gas collection lines were installed, and now they're used when the plant is shut down for maintenance."

Raw landfill gas contains numerous carcinogenic air pollutants, but burning it in a flare, an engine, or a turbine dramatically reduces its overall toxicity. Before the methane collection pipes were in place, Fresh Kills emitted more than fifteen billion cubic feet of greenhouse and carcinogenic gases a year—almost 2 percent of all the world's methane, according to the EPA. Nationwide, landfills are the largest anthropogenic source of methane emissions, accounting for approximately 32 percent of total methane emissions in 2002. Even the best-designed gas collection systems, with wells spaced at about one per acre, at most suck up just 75 percent of emissions, according to G. Fred Lee; that number gradually declines as the equipment deteriorates.

Landfills that collect methane for energy, as well as incinerators that derive energy from waste, receive subsidies in the form of tax credits from the state and federal governments. The subsidies, usually associated with alternative fuels and renewable power, give landfill and incinerator operators a financial incentive to process more waste, and they also make landfilling and burning appear cheaper than recycling and composting, neither of which benefit from tax breaks. Opponents of these subsidies like to remind the public that although landfill gas is an alternative to fossil fuel, landfills themselves are not sustainable, and therefore neither is landfill gas. Their concerns are treated by the waste-hauling industry like so much hot air.

As Diggins drove he muttered a punch list—a drainage swale needed grading, a berm had to be raised. His tour was the engineer's equivalent of a rancher riding his fence lines. At one point I asked about a particularly strong garbage smell. "That's a leachate seep," Diggins said. "There's pressure on the garbage from road material, the stuff we lay down before the final cover. It leaks laterally until enough of it collects, and then gravity pulls it down into the collection system."

Months earlier I'd visited Phil Gleason in his conference room at DSNY's Beaver Street headquarters, and the director of landfill engineering had explained some leachate basics to me. Gleason was portly, with round cheeks, an orange-blond mustache, and a silvery blond comb-over. He had a habit of acting exasperated, of cocking his head and rolling his eyes. Still, it was said that Gleason knew everything there was to know about Fresh Kills, so I had gladly braved his withering glances. For three hours he answered my questions, drew diagrams, and fetched maps to illustrate his points.

"This is a leachate mound," he'd said, drawing a mild bell curve on a sheet of lined paper. "This is what's underneath the garbage. We build a leachate wall around the entire section that goes down seventy feet, to the substrata. Twenty-five feet in from this wall is a perforated pipe. Since the water pressure outside is higher than inside, the only place the leachate can go is in." He said the floor of the landfill was bentonite soil, which was "basically impermeable."

I asked Gleason how William Rathje, in *Rubbish!,* came up

with his figure of a million gallons of leachate daily escaping the landfill's collection system and swirling into the waters that flow along Staten Island's shore. Instantly, Gleason turned apoplectic. "What he did is . . . he made it up! It's not a statistic. Rathje studies trash in Arizona!" Gleason turned his head and looked at me from the corner of his eye. He said, remedially, "Fresh Kills is flushed with hundreds of billions of gallons of water a day. The groundwater here is not used for drinking. The landfill is underlain by clays—it's a composite aquitard!"

Chastened, I turned toward a map and traced with my finger the shoreline I'd paddled so long ago. "That's Carl Alderson's wetlands restoration, isn't it?" I asked. Gleason barked at me, "*His* restoration? *His?* It's ours!"

"Right," I said, and we returned to our cutaway views of methane collection pipes. A couple hours later, I packed up the materials Gleason had laid on me, thanked him for his time, and said, in spontaneous appreciation, that his job seemed pretty interesting. He said, "What I do is boring. All I do is yell at people."

Diggins parked on an elevated roadway so Robin Nagle and I could gaze down upon Plant One, an unloading area in Fresh Kills Creek where a couple dozen empty blue barges were parked cheek by jowl. They looked, from up here, small and manageable. Up close, though, I knew they were cavernous: fifteen feet deep, one hundred and fifty feet long, thirty feet wide, and capable of transporting 1.4 million pounds (the equivalent of four and a half blue whales) apiece. "How many barges are there?" I asked Diggins. "Eighty-two," he said as fondly as if I'd asked how many children he had.

To the west of Plant One, at the intersection of Fresh Kills Creek and the Arthur Kill, was the verdant sweep of the lyrically named Isle of Meadows, a high marsh that had once been farmed for salt hay. After dredge spoils were dumped on the island in the 1940s and '50s, the island's 101 acres of spartina grass and reeds began to give way, along its western flank, to ailanthus, gray birch, poplar, and black cherry, whose seeds could take advantage of this new layer of soil. This upland forest until the past few years provided roosting habitat for six hundred pairs of ibis and heron, plus snowy egrets, black-crowned night herons, and cattle egrets. The insectivorous birds had feasted on flies churned up by the dump's bulldozers. The wading birds had fed on crustaceans that lived in the toxic waters.

With the nesting population growing, Parks Department ecologist Marc Matsil had described the birds as existing "beyond Dante's ninth ring." But then, toward the end of 2001, their population suddenly dropped. Researchers had found cadmium and other persistent pollutants in heron and egret chicks; perhaps it was a simple case of poisoning. Another theory suggested that the birds' historical predators—boat-tailed grackles, red-tailed hawks, and great horned and barn owls—which had sated their appetites on rodents at the landfill until Fresh Kills closed, had now returned to their more traditional prey. The bounty didn't last, and a year after the heron, egret, and ibis populations crashed, those of their predators did, too. It is a scenario that ornithologists are still pondering. Whatever the explanation, the birds were gone.

Nagle and I stood side by side, looking down at Main Creek. I envied her for having visited Fresh Kills while it was going full bore. Since the dump no longer received garbage, I had to mentally walk through the unloading process. It started, Diggins said, when tugs pushed the barges—up to four at a time—in from the Arthur Kill on the rising tide, and a containment boom was closed on the staging area. Out in the Arthur Kill, up to twenty barges could be waiting on deck. Onshore, hydraulic cranes with enormous grapples unloaded, or "dug," one barge at a time, transferring the contents to a concrete pad. The grapples held the equivalent of one fully loaded collection truck, and it took two hours to dig one barge.

I'd seen film clips of what happened next. A front-end loader scooped the mounded trash into a Payhauler, a cartoonishly large dump truck that held about eighty yards of solid waste. The Payhauler trundled up to the active bank. It was as busy as a kindergarten sandbox up there, with dump trucks dumping, dozers pushing loads into two-foot layers, and fourteen-foot-tall, 77,000-pound compactors running over the mess again and again and again. The garbage was buried in fifteen-to-twenty-foot layers called lifts. Fresh Kills had always topped lifts with six inches of soil. At other landfills, lifts received a thinner layer of "alternative daily cover," which could be shredded tires, shredded paper mixed with water, pulverized glass, or foam.

"The slope is three to one," Diggins said, drawing a trapezoid on a piece of paper. "So the height of the bank is twenty feet"—he made a vertical arrow between the plateau and the base—"and the slope measures sixty feet from its bottom edge to the top." He drew another arrow, slanted between these two lines.

How much waste went into one bank? I asked. Diggins squinted at the horizon. "Let's see, " he said. "Six Payhaulers come up from each barge, so that's 480 cubic yards of trash." How much was that in city garbage trucks? We both did the math, coming up with fifteen. "But don't forget," Diggins said. "Banks go on and on. We were building them twenty-four hours a day, six days a week."

As we drove on, I spotted Carl Alderson's small wetland. "Are you doing any wetlands restoration as part of the closure?" I asked, not mentioning that I'd gotten an egret's-eye view of his spartina six months earlier.

"We do shoreline repair," Diggins said, braking to let another truck pass. "We use geotextile fabric and put rip-rap [a jumble of large stones] on the slope."

"Why do you use rip-rap?" I asked, thinking of all the ecological services provided by functioning wetlands—erosion and flood control; filtering of excess nutrients, metals, pesticides, and herbicides; a nursery for fish; a habitat for wildlife.

"Because we want something that's not going to erode," he said, as if it were the punch line to a joke. "We had some people come in to do a restoration," he continued. "They used three kinds of plants, they did 189 of them, and it cost $350,000." (Rip-rap would have cost about a third as much.) Ever polite, Nagle chuckled along with Diggins, but I couldn't join the bonhomie. The new marsh functioned, it was all-natural, and, according to Alderson, it showed no signs of eroding. Even two years on, he wrote me, "it's still the best-looking restoration I've seen in five boroughs and twenty-one New Jersey counties."

We headed toward the leachate treatment plant, a low-rise building in the shadow of Section 1/9 that was currently treating about 800,000 gallons of leachate a day. In his command-and-control office, Don Tuscano, the plant manager, clicked through a series of computer screens that offered schematics of the plant, the landfill, and its underground plumbing. I half expected to see a green horizontal line pulsing like a heartbeat on his home page: the postclosure Fresh Kills, like a brain-dead patient on life support, still throbbed, and its effluents and exhausts were minutely monitored, collected, and analyzed.

"The system we use is called a Sequencing Batch Reactor," Tuscano said to Nagle and me. "Our two concerns are ammonia and suspended solids, like silica and metals. We remove the metals by letting them precipitate out, and we use microorganisms to break the ammonia down into nitrogenous compounds that are then converted to nitrates, which are harmless to the environment."

Nagle was less interested in leachate treatment than in pumping Diggins for information on the Staten Island sanitation garage, which she'd soon be joining (her goal was to work the trucks in every borough). While she and the chief hung back, Tuscano and I strolled outside to the far end of the plant, where leachate flooded from two nine-inch pipes into "influent distribution boxes" made of concrete. The leachate was coffee colored and it produced a bubbly froth several inches thick. "We never get less than two hundred gallons a minute from the landfill," Tuscano said. Flows spiked with rainfall, but the pipes never ran dry, even in a drought: the buried refuse was constantly leaking garbage juice.

I had expected the plant to reek, to smell like a thousand packer trucks on a 90-degree day, but the odor was mild. The solution to pollution is dilution, environmental engineers like to say. It had been the wettest June on record, ideal conditions for visiting a leachate treatment plant. Keith Kloor, writing in *City Limits* about an Alliance landfill in Pennsylvania, described its treatment plant as emitting a "gag-inducing smell—imagine a hog farm overrun with sewage sludge" that "easily penetrates the car, even with the windows rolled up."

From the distribution boxes, the leachate was decanted into holding tanks, then to a mix box, where its pH was raised to 9.5 with the addition of caustic soda. "That gets the metals to coagulate, which makes them easier to remove." Tuscano also added aluminum sulfate, which caused tiny particles to combine and settle, and polymers, which also aided in settling. The leachate rested in octagonal clarifying tanks that had four spokelike weirs radiating from a central point. The liquid in here was tea colored, but Tuscano promised that by the time it was discharged into the Arthur Kill, with its pH adjusted back to normal, about 8.0, the effluent was roughly the same color as Michelob Light. "I could swim in this!" he said in a video shown on cable TV. "What we discharge is cleaner than the Arthur Kill," he pronounced in another video, produced by the DSNY's official artist in residence, Mierle Laderman Ukeles.

The quote-ready Tuscano was quite possibly right, but his comparison didn't count for much: the Arthur Kill was one of the filthiest waterways around. It was a major thoroughfare for tankers and cargo ships, and its banks were lined with chemical plants, scrap yards, oil refineries, and the discharge pipes of municipal sewage treatment plants. I looked at the EPA's two-year Enforcement and Compliance history for Fresh Kills, dated October 23, 2002, and saw that its discharges significantly exceeded federal limits for sulfides and phenols (which came from plastics and were considered hazardous). It also had elevated levels of nitrogen, cyanide, oil and grease, and the outlawed pesticides aldrin, dieldrin, and endrin. And these data covered the previous two years, after the feds ordered Fresh Kills to clean up its discharge.

Tuscano claimed that today's effluent met state and federal regulations. But while he fretted over Fresh Kills' liquid exudation, its solids, the treatment plant's sludge, didn't receive nearly as much attention. It was pumped to a building seventy-five yards away, where it was dewatered and mixed with lime, for stabilization and to raise its pH. Just as sewage treatment plants preferentially moved toxins from water to solids in order to meet the requirements of the Clean Water Act, so did Fresh Kills. It produced thirty-six cubic yards of sludge a day, enough to fill two to three fifty-yard Dumpsters a week. The sludge was delivered to Millville, Pennsylvania, and tipped in a landfill owned by J. P. Mascaro & Sons, which used as its logo a squatting elephant.

Nagle and I ended our tour of Fresh Kills atop Section 1/9, which had received on March 22, 2001, with great media fanfare and a lot of DSNY bunting, the final load of New York City's municipal solid waste. Fifty-three years of landfilling at Fresh Kills had come to an end. Staten Islanders rejoiced, but it was a sad day for Dennis Diggins, who'd made a life here. In September of that year, he was in the process of closing Section 1/9, applying its final layer of cover. "I was right here with my binoculars on top of this mound, watching the towers burn," he said to Nagle and me, looking toward the Manhattan skyline, blurry in the distance. "I knew as soon as they started to fall that the rubble would be coming here. I came down off the hill and got moving. I knew they'd need our equipment."

Debris from the World Trade Center was barged directly from Ground Zero to Plant One, then trucked up to Section 1/9, where police and FBI investigators searched the rubble for human remains. (For a short time, debris was also trucked to the Hamilton Avenue marine transfer station, on the Gowanus Canal, before it was barged over to Fresh Kills.)

The borough of Staten Island lost nearly two hundred people in the tragedy, including firefighters and civilians with jobs in the World Trade Center. The interment of their remains, and the towers' 1.2 million tons of rubble, turned Staten Islanders' opinion of Fresh Kills upside down. The landfill had been the shame of the island: now it was hallowed ground upon which a memorial, with federal funds, would soon be built. Under a proposal introduced by US congressman Vito Fossella, representing Staten Island and

Brooklyn, the 160-acre site would be transferred to the federal government and managed by the National Park Service. Though much has been made by some writers of the majesty of enormous landfills, the particular majesty of Fresh Kills following one horrible gesture was suddenly a lot less literary and a lot more literal. The place was no longer a dump, in the public imagination; it was the final resting place of heroes.

I had lost no one in the disaster, but I could imagine how important this place would be for those who had. Today, the site looked unremarkable—it was a curving field of reddish, stony dirt. There was nothing to identify the ground as sacred. It was in transition: Diggins had to pile on another forty to sixty feet of cover atop the crown to achieve the regulatory geometry. When Section 1/9 was finished, it would be as tall as a twenty-two-story building.

Part Two

Avoiding the Dump

Chapter Five

Behold This Compost

I'd gotten a feeling, by now, for the sordid afterlife of garbage in a dump. But there were branches of my waste stream that never made it to a landfill, that flowed in alternative channels that fanned out, alluvially, into seas of material optimism, the antithesis of a place like Fresh Kills. Some of that optimism was drawn from the promise of recycling, which I'd soon be tracking. But first came optimism drawn from the promise of rot.

Pawing through every single item in my kitchen trash bag to quantify my output, I hit upon a garbage fundamental almost instantly: the worst things we threw out were the things that had once been alive. Organics rendered everything in the can loathsome to touch and smell. By organics, I didn't mean sheaves of lavender lovingly grown without pesticides in the California Floristic Province. Organics, in the garbage world, meant carbon-based stuff that could, theoretically, be transformed into a valuable horticultural product. Food waste was making my trash heavy and wet. Leftover spaghetti coated the plastic bread bags, eggshells dripped albumen on the jelly jars. I was driven to wear rubber gloves, which interfered with recording data.

And then there was the smell. It wouldn't have been a problem if I had put my garbage outside every day, which is what most New Yorkers did. But having to quantify everything slowed things down considerably. Days went by when I didn't have the time, or the will, to process my trash. So it sat around longer, and it made its presence known. Soon, my apartment took on the noxious qualities of a transfer station. It didn't take long for me to realize that if I was going to minimize my garbage footprint—make it lighter and smaller and more pleasant smelling—I had to get the food out of there. I was going to have to compost.

New York City doesn't seem as if it would be a hotbed of composting activity. But once I started poking into the matter, I tapped easily into a vein of composting fanatics. There were home gardeners arguing barrels versus bins; there were community activists at the city's green markets proselytizing for earthworms that devour food scraps in countertop boxes. Park Slope itself was a composting nerve center, virtually preselected for success with backyard and front-yard gardens, a high level of environmental awareness, several organic-food stores, a food co-op, community gardens, and a weekly farmers' market. When the Department of Sanitation started a pilot project ten years ago to collect household food scraps to be composted at Fresh Kills, the neighborhood reached a capture rate of 41 percent.

To help us compost on our own, after the pilot project ended with a round of budget cuts, the Brooklyn Botanic Garden sold composting bins at subsidized prices. My downstairs neighbors Simon and Lori had bought a bin a couple years ago but never assembled it. They were happy to have me set it up in the front yard, right next to the trash cans.

The black plastic Garden Gourmet bin was manufactured in Ontario; its shipping box was made of 100 percent recycled materials; the trilingual instructions promised I'd need no tools. Already, I felt good about this project. I didn't even mind that the side panels, now slightly warped, wouldn't slide easily into a stack. A couple whacks with the hammer and everything was in place.

Everything, that is, except for a square "rodent screen." That was troubling. Did I need it? Where did it go? The booklet seemed to offer no guidance on this matter, though I couldn't be sure. Most of its pages were obscured by splotches of purple, black, and yellow mildew. The decomposition process had already begun.

On a mold-free page, I read the list of things I could and could not compost. Fruit and vegetable peelings, eggshells, tea bags, coffee grounds and filters, bread, pasta, peanut shells, peat moss, garden scraps, wet leaves, fresh lawn clippings, corn cobs, corn husks, clean cotton rags, dryer lint, string, rope, and hair (if it wasn't chemically treated) were in. This "green" material was to be mixed every time we dumped with similar amounts of "brown" material: wood chips, hay, twigs, wood ashes, and dried-out leaves and lawn clippings. Meat, fish, dairy products, and oily and buttery sauces were out. Once I had committed these guidelines to memory, I would tear the booklet into small pieces and toss the scraps inside: the Garden Gourmet would cannibalize them.

I was eager to witness the parade of arthropods that would, in theory at least, arrive on the heels of bacteria, fungi, and actinomycetes (filamentous or rod-shaped microorganisms). The next round of visitors would include white worms, earthworms, sow bugs, land snails, and slugs, followed by beetle mites, mold mites, flatworms, and springtails. The dregs of the dregs would attract ground beetles, centipedes, pseudoscorpions, predatory mites, and ants. Or so the scholars of biodegradation promised.

While I sat in my yard making micro-adjustments to the bin, passersby offered positive feedback. "Composting?"

I'd nod.

"That's so great."

People loved to talk about composting, even if they didn't personally do it. It was a universal feel-good topic—or maybe it was just the crowd I ran with. I went to country houses and chopped up vegetables, and it was natural to ask, "You compost?" Sometimes the answer was yes, sometimes it was no, but always the no came with a regretful shrug that bespoke, perhaps, lost opportunity. Composting, it was understood, put us back in touch with nature's most fundamental processes. It got the smelly stuff out of our kitchens, it kept weight out of the landfill, it nourished our fruits and vegetables. What could be better?

Of course, that was before my bin started overflowing, before it became strongly aromatic, before it was shrouded in fruit flies. For every successful composter, there were probably a half dozen neophyte composters angrily punching the telephone number of the local composting help line. I should have realized, once I started running into officially sanctioned master composters, that tossing food into a bin was hardly the end of it.

But all that was later. Now I lay the rodent screen at the bottom of the bin, though there was nothing to hold it in place. "Installez le couvercle et commencez a composter!" read the final instruction. I positioned the lid and wiggled the bin into the yard's corner, hard by the dryer vent. I wondered if the vent would be a good thing—its warmth might abet decomposition—or a bad thing—it could contribute too much moisture. I raked around our crab apple tree and tipped a layer of leaves into the bin. But where would I get my next load of brown material, and the next? And what about worms, which many said were essential to the composting process?

Across the nation, the number of municipal composting sites rose from 651 to 3,227 between 1989 and 2002. In *BioCycle*'s most recent "State of Garbage in America" report, thirty-five states said that they recycle organics (including yard trimmings and food residuals) and wood. Seattle, which has one-fourteenth the population of New York City but bigger lawns and a longer growing season, produces 47,000 tons of yard waste a year, all of which gets composted. (New Yorkers discard more than 78,000 tons of lawn clippings and leaves a year, and during the so-called leaf season, from fall into early winter, the Department of Sanitation sends special trucks around thirty-five of the city's fifty-nine sanitation districts to pick this stuff up and tip it at a composting area in Fresh Kills. The extra shifts cost the department money, but leaving this debris in the waste stream takes up valuable space in landfills.) Like twenty other states, Washington bans green waste from burial. Los Angeles, with a population of 3.7 million, mixes a quarter of its 300,000 yearly tons of yard trimmings—more than 30 percent of the residential waste stream—with sludge from its wastewater treatment plant to make a fertilizer called TOPGRO; the remainder of the yard waste is composted or mulched.

According to New York's Department of Sanitation, edible and inedible food debris accounts for about 15 percent of household garbage within the city (the second-largest fraction after paper), while the national average is about 9 percent. New Yorkers throw away more food, it is suspected, because most apartments don't have food waste disposers. (The appliances were illegal in the city until 1997, and relatively few households have caught up with the change in regulations.)

Not counting yard waste, my little family was soon dumping an average of 2.18 pounds of organic matter into the compost bin every two and a half days. The weight was mostly coffee grounds; the rest was potato peels, onion skins, and the bread crusts my daughter refuses to eat. (According to the Garbage Project, the average elementary school student throws away three and a half ounces of edible food a day. Over the course of a month, that's equivalent in weight to about three hundred Big Mac foam clamshells, at .233 of an ounce each, to put things in perspective.) I collected this stuff in a small metal bucket near my kitchen trash can. Lori and Simon, on the first floor, and Cynthia and Sid, on the second, began their own composting programs as well. The bin rapidly filled.

My garbage was now light and dry. I was delighted with the change. But about once a week, or more often if we had carnivorous company, things took a turn for the worse. The composter, I realized, was getting the cream of the food waste. The worst stuff—the guts of chickens, the bones of fish—was left behind, for the Garden Gourmet would brook no animal products.

I knew that I could freeze my food waste and lose the smell. But I met a science teacher at a garbage symposium who warned me against it. "The food will become anaerobic as it thaws," he said, sampling a cake shaped like the Fresh Kills landfill, complete with ersatz methane-venting pipes. "You want oxygen in there. You need it for decomposition." He gave me his number and urged me to phone if I had any other questions.

What makes meat so much worse than vegetables? Indeed, why does garbage smell so bad? I phoned the Monell Chemical Senses Center, in Philadelphia, and spoke with Pam Dalton, an experimental psychologist who studies how people are affected by odor. "The putrid smell of garbage is sulfur and nitrogenous compounds," she said. "As waste heats up, a lot of chemical compounds will be liberated by microbes, but the first ones we notice are sulfurous. We have a low threshold for smelling them." With freezing, Dalton said, the gas molecules that carry odor become unavailable. "It's called partitioning, when a chemical changes from one form to another. Over time, and with temperature, the smell becomes more available to the nose."

The rotting flesh of animals produces organic acids that attract flies, which aim to deposit their eggs in a food source that will nourish their growing larvae. "Vultures are also attracted to those sulfur compounds," Dalton said. "That's one of the early signs of decomposing organic matter in the wild. Ethyl mercaptan, a sulfur compound, is added to natural gas to odorize it. When there's a break somewhere in the pipes, linemen locate it by looking for vultures in the air."

I had called Dalton because I'd been told she knew more about how we perceive stinkiness than just about anyone else in America. To profile a smell, Dalton first identified its constituent chemicals. She shot either a liquid or gas sample into the long coiled column of a gas chromatograph. "Chemicals go through phases at different temperatures," she said. "At room temperature you can smell acetone, for example. Other smells take a higher temperature to notice." As a complex mixture was gradually heated in the column, the chromatograph separated its ingredients into individual chemical components that could be detected and quantified by comparing them to a reference standard. So sensitive were her machines, they could pinpoint aromas that occurred in amounts as low as one part per billion. (Still, the human nose—except for the noses of san men, who are inured to stink—betters the machine by more than two orders of magnitude: it can detect aromas at a few parts per trillion.) Among the compounds Dalton had run through her chromatograph were garbage, decayed mice and rats, the famously stinky durian fruit, and urine and feces from many animals, including humans.

These efforts were not blue-sky research. Monell had been approached by the Department of Defense for help in creating a universally offensive odor that could be used for, among other things, crowd control. According to an article in *Chemical & Engineering News,* tests show that putrid odors are "potent in making people want to flee in disgust." They also cause increased heart rate and shallow breathing, and can lead to nausea.

Dalton and her colleagues made chemical mimics of various smells and tested them on people. They discovered that a person's interpretation of an odor depends on his or her upbringing. Aroma and memory are linked. "I had a student from Beijing smell the worst female gingko fruit, and he said it smelled like the cooking odors from the exhaust fans behind Beijing restaurants," Dalton said. "It wasn't offensive to him at all."

The Monell researchers focused on biological odors, Dalton continued, "because we thought those had the best chance of being recognized universally. People really hated these odors." Eventually, the lab came up with a winning nonlethal "odor bomb." Its signature elements were reported to smell like human waste, burning hair, and rotting garbage.

By November, a month after setting up my composter, I had diverted a running total of twenty-six pounds, four ounces of food waste from the landfill in Pennsylvania—the landfill I hadn't yet received permission to visit. By December, I had saved the city another chunk of change by keeping an additional sixteen pounds, twelve ounces from the great waste train. According to the EPA, 67 percent of America's household waste stream could be composted;

I was diverting 37.7 percent of my kitchen's putrescible waste, the stuff I measured on my kitchen scale. So the government and I were out of sync (because the EPA included yard waste and paper in its compost figure), but that was nothing new.

By now, the Garden Gourmet was three-quarters full. We didn't have any actual compost yet, but I believed things were going well. The pile didn't smell, and I didn't notice any vermin around. Still, guilt gnawed at my conscience. I knew that I wasn't always aerating the pile properly. And sometimes if I was in a hurry, or if it was raining, or if I was wearing a skirt (the bin was closely flanked by a boxwood shrub that tore at tights), I didn't even try. I knew that decomposition slowed dramatically in the winter, and I worried that the bin would top out long before the weather warmed up. Would turning my compost more often really help? Did I need more brown material? Did I need worms?

For vermicultural assistance, I considered subscribing to *Worm Digest,* a quarterly out of Eugene, Oregon. I flipped between dozens of worm-related Web sites on the Internet, read meandering threads in composting and master gardener chats. My head swirled with arguments regarding hot versus cold composting, the questionable value of newspaper in compost, whether or not kitchen grease was welcome, and with methods for stemming nitrogen loss, upping carbon-to-nitrogen ratios, suppressing weed seeds and pathogens, nurturing disease-suppressing microbes, avoiding exposure to the elements, and achieving the proper ratio of wet to dry materials.

It seemed so much simpler to throw our food waste to the pigs, the way our urban forebears and our country cousins did. As late as 1892, a hundred thousand pigs roamed New York City's streets, feasting on scraps tossed out doors and windows by the working poor, who relied on these animals to convert waste into edible protein. The pigs weren't docile: they were wild animals that defecated on sidewalks, copulated in public, and injured and occasionally killed children, according to historian Ted Steinberg in *Natural History* magazine. Crusading mayors occasionally passed antipig ordinances, starting in the 1810s, but they didn't stick. Riots to free captive swine (potential pork roasts) broke out in 1825, 1826, 1830, and 1832. "The fatal blow to the urban commons came in 1848," wrote Steinberg. "Cholera broke out in New York, and health officials linked the outbreak to the city's filthy conditions." Club-wielding police herded thousands of pigs from the dwellings of the poor. By 1860, the mayor declared the area below Eighty-sixth Street a pig-free zone: by the 1890s, every last street pig had been slaughtered.

I'd been reading *Charlotte's Web* to Lucy during these confusing early days of composting and was struck by the volume of food little Wilbur devoured. Breakfast: "skim milk, crusts, middlings, bits of doughnuts, wheat cakes with drops of maple syrup sticking to them, potato skins, leftover custard pudding with raisins, and bits of Shredded Wheat." Lunch: "middlings, warm water, apple parings, meat gravy, carrot scrapings, meat scraps, stale hominy, and the wrapper off a package of cheese." Supper: "skim milk, provender, leftover sandwich from Lurvy's lunchbox, prune skins, a morsel of this, a bit of that, fried potatoes, marmalade drippings, a little more of this, a little more of that, a piece of baked apple, a scrap of upside-down cake." Unless Wilbur was depressed and weepy, he rarely left anything for his barn mate Templeton the rat. If I could keep a pig on my roof (the environmentally correct green roof of my dreams, which would function as a personal wastewater treatment plant, a storm-water abatement system, and a microclimate modifier), I felt confident I could achieve Zero Waste, a goal I'd begun to hear vague rumblings about along my trail of trash.

I continued going to garbage events around the city—a roundtable on recycling, city council meetings, garbage art shows, garbage movies—and I kept running into Tom Outerbridge, who ran an outfit called City Green. His firm worked with the city and with the private sector to establish programs and facilities for dealing with city waste, especially the organic kind.

Outerbridge didn't look like the other garbage folk. His shirts were tailored and pressed; his blond hair was lightly gelled and dramatically swept back. He looked more like a TV newsreader than a city employee or a garbage activist, yet he had been director of composting for New York's Department of Sanitation (the position no longer exists) for several years and took credit for launching the composting programs at Fresh Kills and at Rikers Island, a city prison that handled up to forty thousand pounds of food waste a day.

We met for coffee one December morning in Manhattan, and I asked Outerbridge about the origins of city composting. "It started in New York in the early nineties as a collaboration between the Parks Department, which had land that was sort of derelict and needed compost, and the Department of Sanitation, which had a lot of trucks," he told me. But every time the city budget was tight, recycling and waste-prevention programs got cut. Today, the only organic matter the city composted was the leaf-season material and yard waste tipped by the Parks Department and private landscapers at Fresh Kills.

"The economics of composting are challenging," he went on. "Is it cost-effective to collect? In Park Slope, there used to be four trucks going around on one day: the garbage truck; the metal, plastic, and glass truck; the paper truck; and the compost truck. Those trucks have a huge environmental impact. And then there are all the trucks and bulldozers pushing the compost around at the facility—it's very labor intensive."

"Do you compost?" I asked.

"No. I may bring a bag of it to the green market now and then, but I'd rather spend my time working on bigger solutions."

Outerbridge wanted to talk about anaerobic food digesters, his next big cause. There were about seventy-five of these contraptions, which accelerated the composting process, in Europe, a few of which he'd be seeing over his upcoming Christmas holiday. In the United States, Seattle was considering building a digester to handle both pre- and postconsumer food waste from commercial establishments and homes, plus its compostable paper. Digesters were basically enclosed vertical tanks hooked up to dewatering and gas collection systems. "They've got a pretty small footprint,"

Outerbridge said. "You put food waste in one end, usually from some kind of factory, and you cook it until the end product is ninety percent decomposed. Then it spends a couple weeks in an aerobic system, and you get compost. The by-product of an anaerobic digester is forty percent carbon dioxide and sixty percent methane, which you can burn in fuel cells or a microturbine. Or scrub it and get purer methane to use as natural gas. It's a net energy producer."

Outerbridge thought the Bronx's Hunts Point Cooperative Market, through which passed 2.7 billion pounds of produce a year, would make an ideal location for a digester: this single spot, in a single borough of a single city, generated fifty-five tons of food waste and compostable paper a day. According to the EPA, digesting the city's annual output of more than 7 million tons of food and other organic waste, as opposed to burying it, would avoid 1.8 million tons of greenhouse gas emissions and generate 1.4 billion kilowatt-hours of electricity. Outerbridge was prepared to tell me a lot more about food digesters and where they might be stationed, but what I really wanted right now was some free composting advice for my front-yard bin. I told Outerbridge how much green material I was producing every week, that I didn't think there was a lot of decomposition taking place in this weather (in fact, when I was turning my pile the other day, my potato fork had hit solid ice), and that I was afraid the bin would soon overflow.

Outerbridge sighed. "People think composting is the greatest system in the world," he said. "But it takes a lot of energy to make it work—to deal with odor abatement and collecting it and turning it to aerate it."

"Yeah," I said, nodding in sympathy. "But do you think I need worms?"

"No, you don't need worms." He smiled condescendingly. "You need two bins."

I had first learned about red worms many years ago at a green market, where a young German woman sold half-pint containers of *Eisenia foetida,* a common earthworm, for use in countertop composting bins. She also accepted food waste from civilians and turned it into compost on four lots occupied by the Lower East Side Ecology Center, on Manhattan's East Seventh Street. The yard waste and food scraps attracted rats, with which the neighborhood cats could never quite keep up. In 1998, the Ecology Center moved to East River Park, just below the Williamsburg Bridge, and switched from the windrow method, with its long narrow rows of organic material that had to be turned periodically, to in-vessel composting. The process shortened the decomposition time from four months to three by accelerating the rate at which organic matter heated up. It also reduced by about 99 percent the surface area over which rats could forage, and reduced by 100 percent the number of neighbors who might complain about odors.

The name Christina Datz-Romero had come up several times in conversation with city composters, but I realized only when I shook Datz-Romero's calloused hand that the director of the Lower East Side Ecology Center was the green-market worm lady I'd met long ago. She looked a little more weather-beaten now, and she seemed a lot less friendly. The first thing she asked me, in a suspicious tone, was how I'd gotten her contact information. "Your name, your phone number, and your e-mail address are on the Internet," I said. "Anyway, I wanted to talk to you about the composting project because I'm writing a book about garbage—"

"It's not garbage," she interrupted. "It's waste." She smiled coldly.

Datz-Romero sat me down in an old stone building inside East River Park. The sun shone off the water through large windows, and two cats prowled the office, which was sparsely furnished and smelled of disinfectant. "We fished Lucky out of the river in a plastic crate," she said as a black-and-white cat with a harelip boxed with a cord that dangled from my jacket. Lucky's companion, a dull-looking calico, rubbed against Datz-Romero's legs, then against mine.

Here in the park, with the river to the east and greenswards to the west, it was possible to imagine an earthbound community that bonded over seedlings and fresh produce, reveled in seasonal rhythms and cycles completed. Lift your gaze above the park's London plane trees, however, and cookie-cutter high-rises, dozens of them, told another story. "What's the significance of composting in this urban environment?" I asked. Datz-Romero took a deep breath.

"Compost is a natural process, and you can do it from beginning to end. We are community based, and that allows us to be connected and it gives us an understanding of our city as a green and living place. We make our own soil and we plant things. A tree we planted on Seventh Street is now taller than a tenement. You can change your environment and make it more livable. This place is small, but you need to start someplace."

She showed me a desktop worm composter, a round white take-out container with a vent drilled on top. "You could compost the remains of your office lunch," she said. "All you have to do is add worms and damp paper towels." Red worms consume their weight in food every twenty-four hours; their manure, or castings, make superb fertilizer. I savored an image of Anna Wintour, editor in chief of *Vogue,* composting her melba toast crumbs. "It's an educational tool," said Datz-Romero. "It's consciousness raising."

The compost farm was long and skinny: a north-south row of sixteen yard-high grayish cubes with PVC tubes curling out of them. Four days a week, Datz-Romero's truck picked up food scraps from green markets and greengrocers. She blended the scraps with sawdust in the bins, and the PVC pipes, connected to a motor, pulled air through the mixture. From its weekly 3,000 pounds of raw material, the Ecology Center produced 750 pounds of compost. A part-time employee blended the compost with vermiculite, peat moss, perlite, green sand, and black rock phosphate to form New York City Paydirt, which sold for a dollar a pound. I wanted to ask why she had to add so many ingredients if food scraps were such a valuable source of soil nutrients, but Datz-Romero was pushing ahead.

"Leachate collects at the bottom of the air intake pipes and drips through these white pipes to a sump pump," she said, lifting the cover from an orange fifty-gallon pickle barrel. The leachate was frozen, though its molecules were still sufficiently available to remind me of vomit crossed with rotting fruit.

"What do you do with that?" I asked.

"We pour it down the drain and hope it gets treated." In the weeks to come I'd spend some time with inflow and infiltration maps and come to doubt that it did. The storm drain into which she poured about four barrels of concentrated leachate a week was thirty feet from the East River; I was almost certain the drain led straight out.

After four weeks in the gray bins, the food scraps were mixed with some finished compost and moved to a series of wooden curing boxes. Datz-Romero opened one of these and scooped out a handful of lumpy dirt. I saw leek skins and twist ties, seeds, and a whole lemon. Obviously, the compost was at an early stage in its eight-week development, and the worms were not at the top of their game. It was about 19 degrees this morning, not counting the wind coming off the river. Datz-Romero dangled a skinny, stretched-out specimen in the air. "This one is dead," she said. "I'm hoping some adults will survive the winter. They lay their eggs in the fall."

"How do you know it's dead?" I said. It was insensitive of me, but I couldn't help teasing Datz-Romero, whom I found a little tendentious. Moreover, I had by now become slightly disillusioned about worms' alleged power to transform waste into valuable fertilizer. I know that Charles Darwin revered worms: he'd written that they "have played a more important part in the history of the world than most persons would at first suppose." But the creatures have tiny mouths, it turns out, and they are slow and fussy eaters. They won't eat onions, orange rinds, dried-out lemons, or banana peels. They want their food fresh. They won't tolerate a move from indoors to out, and they can't withstand the cold.

Ignoring my question, Datz-Romero led me past a row of sixty curing bins, a blending machine, and a screener, which pulled out the undigested chunks. "Jah," Datz-Romero said, perusing her acreage. "Composting takes a lot of land. You need a partner. We sell most of the finished product in the spring, so we need a place to stockpile it. It's a little smelly. . . ." Indeed, I detected both the nitrogenous compounds from the curing boxes—they weren't so bad—and the sting of leachate from the primary bins.

"Recycling this way is too labor intensive to make money off it," she continued. The Lower East Side Ecology Center had operating expenses of about $100,000 a year. Sales of worms, bins, and compost met 25 percent of that. The rest came from grants. But Datz-Romero had grander schemes. She wanted to leave the small-potatoes world of community composting behind. She dreamed of composting the entire city's food waste in anaerobic digesters. "A digester could take fifty tons a day in a twelve-foot tank and generate methane for energy." It took me a few seconds to realize she was talking about the same project that Outerbridge dreamed of siting at Hunts Point. "The way we compost here, we're creating carbon dioxide, a greenhouse gas. The digester would give us compost, produce energy, and prevent the greenhouse gases of decomposition."

"What about community empowerment?" I asked. "What about changing your environment for the better?" It seemed to me that a digester would shift the emphasis from neighbors who biked their carrot peels to the river and got potting soil in return, to an industrial process that involved massive amounts of materials and trucks. "The first scenario is about a cycle," I said. "The second one is about getting rid of garbage."

"It's not garbage," Datz-Romero said to me, for the second time that day. "It's recycling. It's about making it economically feasible to compost. Our little system is nice and it's loved, but it's not going to be around for too much longer if we can't get sufficient funding. It's an economics of scale—it will generate money through tipping and through selling the end product."

Just before leaving, I decided to ask Datz-Romero about my compost problem.

"Do you think I need worms?" I asked, after laying out the situation.

"It's too cold to start with worms now," she said. "They'd

freeze. What you need to do is cut your food up smaller. The more effort you put into it, the better your compost will be."

I took Datz-Romero's advice to heart. As the air space in my compost bin shrank, I began to put a lot more effort into it. In times of abundance, no one cares about conservation: the beginning of the shampoo bottle always goes faster than the end. When Europeans first explored this country they saw unlimited timber, fish, game, and clean water. Now that our population is so large—or, more accurately, that our perceived needs are so large—we realize that our natural capital is finite. And yet we seem helpless to stop ourselves from taking.

Trying to stretch out the last couple inches in the composting bin, I began to slice my banana peels into squares the size of Wheat Thins and whittle my celery stalks into matchsticks. I wanted more surface area available to hungry microbes. This was tedious and annoying and time-consuming, and I sometimes found myself slashing away with a knife inside the metal bucket. "How can I grind up all this food into little pieces more easily and more safely?" I asked myself. And then a light bulb blinked on over my head.

Food waste disposers were invented in 1935, but they weren't commercially developed until the country recovered from World War II. In 1948, the American Public Health Association predicted that the garbage disposer would send the garbage can the way of the privy, making it anachronistic. (Perhaps they should have said disposers would send garbage cans the way of the pig, which was, funnily enough, the name many householders used for the machine underneath the sink.) But there were holdouts. I, for one, had a strong presumption against disposers. There was the extra water they used, and electricity. But mostly I worried about all that potential food energy running down the drain. I thought Tom Outerbridge, Mr. Compost, would agree.

"Disposers?" he crowed when I asked. "I think they're great!" I was taken aback, then he qualified his statement. "It's dependent on the wastewater treatment system. If it's clean sludge and the system can handle it, so there's no overflow into waterways, then it's good. It adds nutrients to the composted sludge. If it's going to agricultural land, versus picking up food waste on a truck and composting it, the former is preferable. You don't have all those trucks pushing it around."

I asked Christina Datz-Romero for her opinion. "Food waste disposers suck. It's a waste of water. And why would you want to overburden the wastewater plants?" Indeed, the commissioner of New York City's Department of Environmental Protection publicly grumbled about clogging the already fragile system and about harming aquatic life with the increased nitrogen loads.

"But disposer studies say water use is negligible," I said to Datz-Romero.

"Look at who paid for those studies."

I did, and the answer brought me to Kendall Christiansen, co-chair of the Citywide Recycling Advisory Board, or CRAB, and PR flack for the Plumbing Foundation. Writing on behalf of the plumbers, not the city, Christiansen sent me an e-mail:

> *Most environmental issues are trade-offs: in this case, food waste collected by sewer system uses a) high water content of food waste, b) gravity/energy of existing infrastructure to collect vs. diesel-spewing trucks, c) beneficial end-use of all of city's sludge for land application, vs. burying in landfills where it generates methane and contributes to global warming.*

This gave me pause: was my compost bin altering the climate?

Christiansen acknowledged that the treatment plants would have slightly higher costs, but these were offset, he wrote in his virgule-heavy style, by running fewer trucks, lowering garbage exports, and lessening landfill impacts.

> *Given that 2/3 of NYC's residents live in apartments, etc., composting—while good/fine—isn't practical/feasible/likely as reasonable alternate for diverting significant*

quantities of food waste, either short- or long-term. . . . Not suggesting either/or; but where composting makes better sense is with larger-scale generators so that collection/transport efficiencies are more significant.

The studies that Christiansen touted said that food waste disposers used about a gallon of water a day, but critics considered that a gross underestimate.

The ReSource Institute for Low Entropy Systems, which promotes the use of composting toilets and offers a multipoint plan for protecting clean water (it includes unhooking toilets from municipal sewers in favor of composting models), has campaigned to abolish garbage grinders. "It is as irrational to use water to transport food wastes as it is to use water to transport human excreta or industrial wastes," the foundation believes. "Water should be used only for drinking and for washing."

William McDonough, author with Michael Braungart of *Cradle to Cradle: Remaking the Way We Make Things,* bases his entire design philosophy on the notion that waste equals food—that the end of one product's usefulness should nurture the birth of another. McDonough and Braungart break waste into three categories. "Consumables" are things we eat or use that would eventually biodegrade, including shampoo bottles made of beets and fabrics free of toxins, mutagens, and endocrine disruptors. "Durables," like TVs and cars, would be returned to their manufacturers as technical nutrients and used as food in their manufacturing systems. "Unmarketables," like nuclear waste, dioxin, and chromium-tanned leather, would no longer be produced or sold.

Mulling over the waste streams that left my house, I saw that sometimes they intersected—as food and sewage did at the wastewater treatment plant—and sometimes they came apart—at a recycling facility, for example. The line between compost and the majority of material at a landfill is all too thin. City planners on Nantucket, faced with a leaking landfill that threatened marshes, wildlife, and the island's aquifer, recognized this years ago. Decreeing that they would brook no net gain of landfill space, island residents approved installation of a 185-foot-long horizontal digester. Household waste is loaded into the first of five compartments in a rotating steel tube. (Recyclables, including mattresses and textiles, tires, white goods, furniture, and shoes, have already been harvested.) In two to three days, the temperature of the organic material reaches 160 degrees and the remaining nonorganics—plastic bags and the like—are removed. The rough compost is cured indoors for twenty-one days on a ventilated floor, then moved outdoors, screened, and mixed with chipped brush and yard waste to cure for another three months. Waste Options, which runs the program, claims that the island's diversion rate, the amount of garbage that isn't getting buried, is 86 percent.

To handle the remaining inorganic fraction of household waste, Waste Options built a new state-of-the art landfill, complete with a double liner. To meet the island's no-net-gain-in-landfilling mandate, the company plans to mine one cubic yard of the island's leaky old landfill, and run it through the digester, for every cubic yard it puts into the new one. Eventually, the much-reduced old landfill will be capped and covered with fresh compost, reclaimed not for strip malls or condos, but for native plants and animals.

If food waste disposers represented cleanliness and modernity, my own system continued to point in the exact opposite direction. Every few days, I dumped my kitchen trash onto my daughter's blue plastic toboggan, squatted next to it on the floor, and weighed its components on a kitchen scale that, afterward, I barely managed to wipe, let alone "sanitize." The process felt primitive, Luddite even, and I liked that. Not for me the tyranny of modernity, the hysteria of hygiene. Picking through my trash felt subversive: it ran counter to the media message that household dirt should be whisked quickly into a compactor or garbage pail lined with a lemon-scented bag, preferably via single-use mops, disposable dust cloths, and paper towels, which would protect my family from the germs that festered in my kitchen sponge. Composting my organic matter, reclaiming my own mess, was beginning to feel political.

One morning, eight months after I began composting, I went

outside with a small container, hoping to collect some potting soil from the bottom of the compost bin. I'd never opened the sliding hatch before, so I had no idea what was happening at the business end of this thing. As gently as possible, I lifted the door, expecting the cascade of "rich, dark, soil-like material" promised by the Garden Gourmet booklet. What I saw instead was a wall of compressed brown crab apple leaves. I poked around with my finger, trying to see if there was any loamy soil, and ran instead into a small gray mouse. It scampered toward my hand, faked right, and dove left, back into the leaves. That was a surprise, though it shouldn't have been. The rodent screen wasn't really attached to anything.

I considered digging around with the potato fork to see what else was going on in the bin, but I didn't want to impale the mouse or its inevitable relatives. The EPA has a regulation, called 40 CFR, Part 503.33, concerning "vector attraction reduction" in soil enhancements. Obviously, I was out of compliance here. The mouse could potentially transmit disease, but I couldn't help giving my little vector the benefit of the doubt. He was cute, and vegetarian. Setting aside the potato fork, and noting with admiration the dryer lint with which the mouse had sealed his hidey-hole, I slid his front door closed.

In a few more months, my chest-high compost bin would brim with the organic castoffs of my family, my neighbors, their garden, and the tree in our front yard. The volume was a reminder that while packaging makes up 35 percent of household waste in the United States (by weight), yard waste and food scraps make up another 25 percent. The fraction hasn't always been so high. Between 1920 and 1990, the volume of organic matter in the American residential waste stream rose fourfold. Some of that increase was due to population growth, some to increased consumption, and some to more efficient collection. But the expansion came also, writes Susan Strasser in *Waste and Want,* from "a new willingness to define leftover food . . . as unwanted."

Ironically, although we yearly bury or burn fifty million tons of energy-rich yard waste and food scraps (contributing significantly to the production of greenhouse gases) we continue to pump fossil fuel from the ground to produce commercial fertilizer. According to Richard Manning, in *Against the Grain: How Agriculture Has Hijacked Civilization,* it takes an average of five and a half gallons of fossil energy, converted to fertilizer, to restore a year's worth of lost fertility to an acre of eroded land. Nitrogen from this fertilizer (but not from compost, which releases its nutrients slowly) leaches into and pollutes water; the "dead zone" in the Gulf of Mexico, for example, has been attributed, in part, to nitrogen runoff. Every single calorie we eat, writes Manning, "is backed by at least a calorie of oil, more like ten."

Is it possible to take the energy from an urban population's green waste and return it, safely and efficiently, to the earth? In 1895, Colonel George E. Waring found that transforming New York City's organic waste to fertilizer and grease was easy enough, if you had a cheap and willing labor force and neighbors who didn't protest the stench. But getting residents to separate this stuff, and hang on to it until the collection truck came 'round, was an uphill battle. The city's more recent experiment with kitchen scraps and garbage trucks revealed a similar reluctance to separate, at least in multifamily dwellings. San Francisco collects and composts organics (which include oily pizza boxes, greasy waxed-paper bags, and Chinese food cartons), but mostly from restaurants and smaller residences. In Oregon, the Portland International Airport composts food scraps, but it is a tiny and self-contained operation. Seattle, for all its green consciousness, green markets, and green lawns, does not compost residential food scraps, though it is planning to start a commercial program. Europe, as usual, is more enlightened: in the Netherlands and in Germany, residents of single- and multifamily dwellings routinely scrape bread crusts and orange peels into bins for municipal collection and composting.

Because food waste presents a different set of recycling challenges than metal cans and dry newspapers, solid-waste managers are keen on pilot projects that test different types of containers and collection schedules. One size doesn't fit all. Residents in Toronto and Hamilton, Ontario, are currently experimenting with food collection, as are five neighborhoods around Minneapolis and St. Paul. It is obvious to me that front-yard composting bins aren't going to work for the majority of New Yorkers, but that doesn't mean that our food waste—especially the food waste from large institutions and processors—can't be composted. All it would take is political will and the resources to make it manifest. Someday, I suspect, there is going to be gold in green waste. But not today.

Chapter Six

Forward into the Flexo Nip

In 1967, the city of Madison, Wisconsin, was running out of dump space. To stem the flow of garbage, the city hooked special racks onto its garbage trucks and sent them driveway to driveway, collecting newspapers from residents. It was the nation's first curbside recycling program since the forties, when communities had collected material for the war effort. In 1973, the city of Marblehead, Massachusetts, launched its own curbside program, which included bottles and cans in addition to newspapers. The program had been inspired by Earth Day, inaugurated by the League of Women Voters, and was funded, in part, by the nation's first EPA recycling grant.

The curbside habit started to spread, but not quickly. By the late eighties, there were still just a few hundred programs across the nation, but garbage consciousness had reached a critical mass. Teachers had begun writing recycling curriculum, and the media hyped the plight of the barge *Mobro,* which wandered the high seas for nearly two months trying to offload three thousand tons of Long Island's garbage. In 1988, the EPA announced a goal to recycle 25 percent of municipal trash, and the following year Andie MacDowell, in the opening scene of *Sex, Lies, and Videotape,* nattered to her shrink about a crisis in landfill capacity. By 2003, there were more than nine thousand curbside recycling programs across the country.

According to "The State of Garbage in America" for 2003, western states had the highest regional recycling rate (38 percent), and the Rocky Mountain states had the lowest (9 percent, with Colorado at a dismal 2.8 percent). The top states were Maine, Oregon, Minnesota, Iowa, and California, but they seemed to compete on uneven ground, calculating diversion from the landfill using different criteria. Should the stuff rejected by recyclers count as being recycled? Should measurements be taken in volume or weight? Should commercial construction and demolition debris, which was diverted from the landfill, be included with the residential figures?

New Yorkers began recycling paper on a voluntary basis in 1986; it became mandatory with passage of the city's first recycling law, in 1989. From the start, the city relied on a "dual stream" system: one truck picks up metal, glass, and plastic; another truck handles paper. Most US cities use this system, though "single stream," in which all recyclables are thrown into one container and sorted at a MRF, is now getting play in San Francisco, Phoenix, Denver, Los Angeles, and Palm Beach County, Florida. This system cuts truck traffic in half and, because it makes recycling into a no-brainer for citizens confused by chasing arrows and the material properties of juice cartons, could ultimately net more stuff for processors to sell. On the other hand, mashing all kinds of materials together into one barrel, then compressing them in the back of a truck, can take a toll on quality: the paper gets dirty, say the paper people, and the plastic start to glitter with glass shards.

Seven days a week, I set aside paper for the weekly pickup: newspapers, junk mail, and, if I could get away with it, draft versions of Lucy's artistic masterworks. My magazines went on the slippery pile, along with my Band-Aid wrappers and shopping lists, receipts and tea bag tags, toilet paper tubes, egg cartons, rinsed out ice-cream containers, and boxes from Belgian waffle mix, Mr. Bubble, baking soda, and butter. It added up to at least twelve pounds a week, and every bit of it earned the city money.

Unsolicited take-out menus and advertising circulars are an almost daily fact of life in New York, and every time a hired hand dropped another one on my stoop, I heard in my head Judy Goodstein's Queens-accented cackle: "Keep the junk mail coming. Ha-ha-ha. If you can tear it, we can take it." Goodstein was the recycling manager of Visy Paper, on Staten Island. The company processed more than 180,000 tons of New York's residential and commercial paper a year, and it was looking for ways to capture an even higher percentage of what she called "the urban forest," a phrase the company had actually trademarked.

With her heavy makeup and shag hairdo, Goodstein was a fixture on the garbage scene. She cultivated a hapless mother-hen personality. "Do you know what this meeting is about?" she'd whispered to me on more than one occasion. "I hope I don't have to speak." But this reticence was a put-on: Goodstein wasn't the least bit shy. In fact, she was eager to reel off the company boilerplate: how much paper Visy took in, how much corrugated (she pronounced it "carra-gated") linerboard it produced, and how many trees all this saved. Depending on the audience, she would plaintively express her hope that the Department of Sanitation not abandon dual-stream collection. Paper bundled inside plastic bags was bad enough (the city allowed bags because they were easier than tying up paper with string), but glass and metal in the mix would be hell.

My paper arrived at the Visy plant crammed inside DSNY truck CN231, which was loaded on Wednesday mornings by Jack McLean and Jorge Toro. McLean was a muscular black man from North Carolina, a former printer who liked being outdoors, working without a boss, and meeting new people. Toro, who pronounced his first name "George," was a worn-looking Puerto Rican, a former corrections officer from Brooklyn who didn't seem to like much of anything. McLean, known around the garage as Clever Jack, had eleven years on the job; Toro had four, though his relative inexperience didn't stop him from acting as a sanitation sociologist.

"We get between ten and eleven tons of paper on an average day," Toro told me one bright morning as he dragged paper-filled bags and cans from the curb to his truck. "But look at this! It's wet from rain. It's heavy. They're supposed to keep the tops on the cans. People use rubber bands on thirty pounds of paper—in affluent neighborhoods! Yeah, even over on President Street, where the professionals live. On windy days, oh, that's the worst. The paper blows right out of your truck. I don't like new phone book time. Some people save them and throw out a hundred pounds of phone books at once."

He stopped to snatch a crumpled cigarette pack from the gutter. How exemplary, I thought. But no, Toro was only after the coupon inside the cellophane. "You can get some pretty neat stuff with these," he said. "I once got a camera."

He dropped the rest of the pack into an empty garbage can and continued parsing neighborhood compliance. "In Bed-Stuy, Bushwick, Borough Park—there's no recycling over there. They don't know how to do it! The people in Borough Park bring their garbage back from the Jewish Alps, the Catskills, where they go on vacation. Then they complain about us not taking things, and we file a complaint, and the Jewish mayor gets them off."

There was no stopping Toro. "The Chinese are the Jews of the Orient. They compact their trash till it's tiny and they can fill up a can with a hundred pounds. The more money people make, the less they care. I'm talkin' renters—they pay a lot; they want the landlord to take care of things. If people just did as they were told, it would be easier. They put it all in plastic bags, and they overfill them, and then they break."

It was a little after eight now, and Billy Murphy and John Sullivan, who'd whisked away my putrescible waste about two hours ago, had caught up to us on Tenth Street. I yanked my containers a little faster, feeling pressured by one of the fastest teams in the garage, but McLean and Toro wouldn't be hurried. They dragged, they stopped to chat, they walked from house to house.

"They are slow, aren't they?" Sullivan said to me, shaking his head and smiling. He put a foot up on his dashboard and took a slug of water. CN191 couldn't squeeze past CN231, so Murphy and Sullivan resigned themselves to an unscheduled break. When the paper truck got to the avenue, it pulled left. The garbage truck went right.

We were on Thirteenth Street now, where my paper collectors stalwartly ignored a car honking to pass. "If we let one person through, then ten will want to get through," Toro said. "It don't bother me. You see them lights?" He nodded toward the flashing yellow and red lights high over his truck's hopper. "They can see them a block away. They don't have to drive down here."

We picked our way past whitewashed town houses and geranium-bedecked brownstones. In front of a yoga studio, a contractor had double-parked a van. "Same thing every week," snorted Toro. "This guy is supposed to be so holy." He gave three long blasts on his horn, enough to pull a wizened yogi from inside, his robe and white beard flowing. "What do you mean?" I asked Toro, not making the connection between the contractor's van, the yoga studio, and the snort. "It's a tax write-off for him, a religious organization," Toro barked. I risked having my head bit off, but I told Toro his truck could probably fit past the van. "I don't want to risk it," he said piously. "If someone hits us, we have to go to court. You lose a day's pay." That, and it was way more fun to make a yogi run.

The Visy Paper plant occupied the ur-edge of town, a desolate spot on the sort of exurban borderland romanticized by numerous Bruce Springsteen songs. The potholed boulevard that ran past the factory's front gate dead-ended in the marshes of Staten Island's western shoulder. At high tide, water crept between cattails to lap at what had become a visual convention for illicit pleasure: shattered beer bottles and spent condoms. A couple hundred yards away, steam from Visy's smokestacks wafted over fields of phragmites and dissipated over the West Shore Expressway.

The factory itself was a utilitarian metal box with minimal landscaping. A bright mural out front illustrated various corporate talismans: green trees, blue skies, and the New York City skyline, complete with twin towers. I found one element of the mural cryptic—a ball field upon which stood a brown-skinned man in coveralls, arms stretched wide, feet and chest bare, his head tipped rapturously back toward puffy clouds. What did it mean?

Judy Goodstein, when we met in her office, didn't have a clue. I was here for a tour, to see how 360,000 tons of paper—nearly half of it collected by the DSNY—was turned into the equivalent of about 650,000 new cardboard boxes a year. (Not all of New York's wastepaper ended up at Visy. The city had contracts with a handful of independent paper dealers, some of whom sold material to Visy, or sorted and baled it for sale to other recyclers, or shipped it overseas, mostly to Asia. Wastepaper was America's largest export, by volume, to other countries. Where the city's paper went depended on the market, and the market was highly variable. Visy also bought commercial paper from private carters, who collected 170,000 tons a year from offices and institutions.)

Goodstein pressed upon me a pair of foam earplugs and led me into the bowels of the plant. Making paper is a noisy, dirty process, but my tour started at the concluding end of the mechanical mayhem, where a single 200-inch-wide roll of brown paper hung serenely from the ceiling. At this point, Goodstein handed me over to production supervisor Chris Lovett, who whisked me straight back to the beginning.

A quarter of Visy's residential paper supply arrived aboard DSNY trucks that plied parts of Brooklyn and all of Staten Island; the rest arrived on barges loaded, from packer trucks, at a marine transfer station on Manhattan's West Side. Both supply lines ended in a thirty-foot-deep pit on the plant's western edge. Peering over a railing into this concrete hole, I saw examples of nearly half the sixty-odd paper types listed on the Loose Waste Paper Exchange, a price index published by trade groups, including boxboard, beer carton waste, white envelopes, colored ledger, hard white, soft white, glassine, brown kraft, and "other."

The first step in papermaking, Lovett impressed upon me, was paper cleaning. It started when the loose magazines and the junk mail, the boxes and the paperboard, were transferred, via a grapple the size of a small shed, from the pit to the pulper, which sloshed the stuff around with water to separate wood from plastic fibers. "We skim the big stuff off, then pump it through another cleaner that takes out plastic, staples, tape, glue, and clay, which comes from colored paper." In the DAF, or Dissolved Air Flotation unit, Lovett continued, air surrounded contaminants and floated them to the surface for skimming. We walked along a high iron catwalk, where Lovett directed my gaze at a vat twenty feet in diameter. A shroud of steam obscured its contents until a sudden draft revealed a surface of bubbling brown scum: primordial paper soup.

After the initial cleaning, the slurry went into a forty-foot dump chest, where it was again mechanically agitated. "It's continuous cleaning throughout the plant," Lovett said. "We're basically just wetting and rinsing here." He showed me an outdoor waste pile that contained rock, sand, glass, popcorn, and the shredded remains of the plastic bags that McLean and Toro hated to see. In an instant I realized that putting waxed-paper bags in my recycling pile wasn't the worst thing in the world, that I didn't need to tear out glassine windows from pasta boxes or get every last popcorn kernel from the microwave bag. Chris Lovett would do it for me.

Back inside, Lovett pointed to another machine churning out a thin brown crumbly material—more rejects. "It looks like potting medium," I said. Pleased, Lovett said I was correct: the company hoped to sell it to gardeners. Paper recycling mills, in terms of waste, are hardly zero-sum operations. They produce far more short-fiber sludge than do virgin mills. In fact, 15 percent or more of the incoming paper comes out as sludge. Contaminated with bits of plastic and strapping, it makes poor compost. Visy landfilled its dirty remains but entertained notions of drying and burning them to power the mill. The company would save on fuel and on trucking to a landfill. The idea of a new smokestack on the horizon, however, didn't sit well with the Staten Islanders who had fought incinerators in their borough for twenty years, and ultimately the mayor nixed the plan.

After four centrifugal cleaning sessions and a bit more dewatering, the slurry was forced through hoses and sprayed onto a screen that fractioned, or sorted, the long fibers from the short. Lovett wanted the long ones on top of the sheet because they looked better and they were stronger. Visy received a lot of ONP (old newspapers) and MOW (mixed office waste), which had short fibers. Long-fibered OCC (old corrugated cardboard) vastly improved the quality of the mix, and Goodstein was perpetually on the hunt for more of it.

She also preferred fancy office paper to the 100 percent postconsumer stuff that I regularly bought, used, and then shunted to the recycling pile. (Postconsumer fiber is paper collected through recycling programs; preconsumer fiber is mill trimmings and scraps discarded on the manufacturing end. The higher the postconsumer content, the more resources are conserved and the more waste and pollution are minimized.) As paper is recycled again and again, its fibers become shorter and shorter and it becomes useful only for lower grades of paper, like the linerboard used in cereal and shoe boxes, or paper towels. According to the giant paper manufacturer Weyerhauser, clean white paper can theoretically be recycled nine times, but the reality of inks, clays, and glues drags that number down to four times. When the fibers become too short to be rewoven into paper, they are washed out into the reject pile, to be reincarnated as kitty litter or, in Visy's case, potting soil.

I returned my attention to the chugging machinery. At this point in the recycling process, the paper was more like watery gruel than something you could fold into an airplane: it was only 8 percent solids. A dewatering machine brought the density up to 14 percent, and then an apparatus called a Kramer, made up of long rollers, thickened the pulp to 28 percent through a process of vacuuming and pressing.

Cleaned and reorganized at the structural level, the slurry rolled on. "It comes in here at about one percent solid," Lovett said, leading me through a door that separated the subtropical region of the plant from the tropical, "then gets up to twenty-one percent within about thirty feet." A "Flexo Nip" squeezed water from the paper, bringing it to 50 percent, and then it moved in one continuous ribbon to a series of drying cells. Lovett lifted a roll-up wall so I could see the paper flowing in a mile-long serpentine pattern around fifty-seven cylinders. A blast of heat hit my face, and I quickly stepped back.

"We add cornstarch to the paper to strengthen it, which brings it back to twelve percent wetness, and then dry it back down to nine percent," Lovett continued. "Then we smooth the top for printability, that's called calendaring, and add dye so it's the color of paper, not gray. People don't want that."

Three hours after my ONP and MOW had tumbled into the pit, they reemerged from the machine as brown linerboard, tightly rolled onto a spool twelve feet in diameter and then sliced by giant knives into something your basic forklift could manage.

Seventy percent of the paper produced at the Visy plant on Staten Island was fed to a Visy plant in Valparaiso, Indiana, that produced boxes made of 100 percent recycled fibers and linerboard, and the rest of the paper went to other Visy box plants and corrugators in the Northeast. Goodstein liked to call the environmental savings of recycled paper "the TOWEL effect"—for trees, oil, water, energy, and landfill. "We benefit all of that," she said, meaning that she saved trees from being turned into pulp (the Staten Island plant claimed to save 13,500 trees a day), saved oil burned in logging operations, saved water by recycling 600,000 of the 700,000 daily gallons used inside the plant, used 25 percent of the electricity consumed in a virgin wood–pulping operation, and kept 350,000 tons of paper a year out of landfills. Actually, about 15 percent of the weight Visy received had to be landfilled—dirt and plastic and the other stuff agitated out of the mix. But virgin mills were even worse: 25 percent of a harvested tree went onto the waste pile.

Virgin papermaking is one of the most environmentally harmful industries on earth. It depletes forests and their biodiversity, it uses more water than any other industrial process in the nation (more than double the amount of recycled papermaking), and it dumps billions of gallons of water contaminated with chlorinated dioxin and a host of other hazardous and conventional pollutants into rivers, lakes, and harbors. According to the Natural Resources Defense Council (NRDC), the paper industry is, after chemical and steel manufacturing, the third-largest source of greenhouse gases in the United States. Each year, paper factories send 420 million metric tons of carbon dioxide, water vapor, methane, nitrogen oxides, and other heat-trapping gases up their smokestacks (and emissions are expected to double by 2020). Along with the gases come 38,617 pounds of lead and 2,277 pounds of mercury and mercury compounds. The mercury, released by plants' coal-fired boilers, settles in water, where bacteria transform it into a highly toxic form called methylmercury. Small organisms, like plankton, consume the methylmercury and are in turn consumed by small fish. The small fish are eaten by larger fish, which are in turn consumed by other animals, like us. Ingest too much methylmercury and your kidneys and brain are ready to reenter the nitrogen cycle.

Converting old paper into new paper avoids many of these environmental costs and, by creating jobs and transforming a material worth $10 per ton to paper collectors into a material worth in excess of $400 per ton to linerboard buyers, makes sound economic sense as well. Recycled newsprint, which goes for $600 to $700 a ton, is an even better deal.

While residential and commercial paper-recycling rates across the United States have steadily increased—from 30 percent in 1988, when the American Forest & Paper Association started to keep track, to 50.3 percent in 2002—consumption of virgin paper has steadily risen as well. Over the past fifty years, according to the independent market research firm Nima Hunter, worldwide use of virgin paper has increased sixfold, with the average US office worker running through more than ten thousand sheets of printing and copying paper per year. Ninety-five percent of the twelve billion magazines printed annually in the United States have zero recycled content, and only about 20 percent are recycled, which puts more than nine billion magazines into landfills and incinerators every year. According to the Worldwatch Institute, virgin wood pulp consumption continues to expand at roughly 1 to 2 percent a year, making recycling efforts all the more crucial.

"We've got to keep doing it, even if we're just holding the line," the NRDC's Allen Hershkowitz told me. "Recycling paper *is* slowing global deforestation. Conservatively, timber harvests would expand fifty percent in the next thirty-five years if we didn't recycle paper."

We need to make sure we are *buying* recycled, too, because shifting all this household paper from the landfill to a company that pulps, dries, and flattens it isn't worth a thing if economic markets don't signal its value. Collectively, the United States consumes more than eighty million tons of paper a year, and less than a third of that comes from recycled sources. The paper pushers of Washington, D.C., the nation's largest buyer of office supplies, are required by an executive order to purchase products made with at least 30 percent postconsumer recycled content, when available. Any state or local procurement agency that uses federal funds has to do the same. Buying recycled products ups the demand for more recycled products, which in turn saves even more resources, reduces pollution, and lowers prices for those goods.

A solid-waste expert, Hershkowitz had the intensely focused, suffer-no-fools manner of a Beltway policy wonk. He angled forward in his chair, the better to spring up and snatch documents from his several shelves of garbage-related reports. His walls were covered with pictures of his children; a framed *New York Times* review of his book, *Bronx Ecology: Blueprint for a New Environmentalism;* and an actual blueprint of the book's subject, a recycled-newsprint mill he had conceived of, and designed with the architect Maya Lin. The mill would have employed residents of Hunts Point, recycled gray water from a nearby sewage treatment plant, and kept bargeloads of paper from landfills (as well as a few from Visy, with which it would have competed for commercial paper). After nearly ten years of negotiations and delay, Hershkowitz's dream was crushed by lack of support from City Hall.

I didn't find it strange that Hershkowitz had a roll of toilet paper on his office shelf, and he didn't find it strange when I pulled a toilet paper wrapper from my backpack.

"What do you think of this?" I said, showing him the high postconsumer recycled content of my brand, Marcal.

"Pretty good, very good," he said.

Why, then, didn't Marcal advertise its green credentials, like Seventh Generation (which cost nine cents more, even at my food co-op, for a roll half as long)? "It's a public perception thing," Hershkowitz said. "People care a great deal about what they use on their bottom. Even though it's there for only five seconds. I timed it."

"If it's recycled, some people feel its quality is less or that it's less clean," said Peter Marcalus, senior vice president for recovered materials at Marcal, in New Jersey, when I phoned him later. "I've had countless people at public meetings say to me, 'I just can't understand how you get my used tissue clean again.'"

Hershkowitz was currently knee-deep in a campaign to halt the pulping of Canada's boreal forests—a vast, ancient, and globally rare ecosystem of hardwoods and high mammalian biodiversity—for the creation of toilet paper. To some, this was an environmental crime on par with converting wetlands to shopping malls. It was toilet paper that had brought me to Hershkowitz in the first place. It had been he who'd answered a reader's query posed in the the *New York Times*'s science section: "Is it better for the environment to dispose of toilet paper in the garbage or in the toilet?" It was a question I had asked myself for years, every time I blew my nose.

Hershkowitz had answered: "It is both practically desirable and ecologically superior to flush it down the toilet." Sewage treatment plants decompose both organic waste and the cellulose of toilet paper, he had said, but in a landfill, microbial activity would generate methane, a potent greenhouse gas. No matter where it ends up, though, the big ecological problem with toilet paper, Hershkowitz pointed out, is that companies produce it from virgin wood instead of recycled paper. If toilet paper is made directly from trees, the yield is only 43 to 47 percent (depending on what type of wood is used). But make toilet paper from wastepaper, and the fiber yield is considerably higher: between 85 and 95 percent.

"Paper recycling does save water, energy, and forests," Hershkowitz said, "but because the extraction of virgin timber is subsidized by the government, it's still cheaper than buying old paper." The paper subsidy favored well-financed and politically influential extractive industries over recycling and reuse enterprises. Where the former tended to deplete natural resources, pollute the air and water, and eliminate jobs, the latter tended to be resource efficient, entrepreneurial, and community based. According to a 1999 report by the GrassRoots Recycling Network, Taxpayers for Common Sense, the Materials Efficiency Project, and Friends of the Earth, tax breaks and spending subsidies for timber, mining, and energy extraction—which were instituted in the nineteenth century to help develop the West and industrialize the nation—cost taxpayers more than $2.6 billion a year. Eliminating the subsidies would "give recycling and reuse industries a more even playing field on which to compete while also saving taxpayer money." Not surprisingly, politicians accustomed to donations from extractive industries have no interest in changing the situation.

On a wintry morning I walked up the ramp of the Department of Sanitation's marine transfer station on Manhattan's West Side and gazed down from a platform upon the swirling black Hudson. According to the office blackboard, the station was home to the Fifty-ninth Street Killer Whales, who were "Sleek, Powerful and Fearless." The Killer Whales, who in their green sweatshirts and trousers looked like any other DSNY employees, were a study in energy conservation today, minimally supervising paper-filled packer trucks as they backed their loads to the edge of an elevated tipping platform. Far below waited an enormous blue barge, of the same size and type that once carried putrescible waste to Fresh Kills. A single barge could hold 450 tons of paper, and when four of them topped out, the Killer Whales would fearlessly cover them with netting, sleekly lash them together, and then hand them over to a tug, which would push them down the river and across the harbor to Visy.

The metal building on the one-acre pier was as cavernous as an airplane hangar and sheared by river winds. Nor'easters and high tides had been known to trap barges against the platform's underside; ice locked tugs and barges in place. Dust swirled in the summer heat, combatted by san men wielding hoses. "Men have fallen in, but no one has drowned," Austin Johnson, the shift supervisor, told me.

From the lip of the platform, I watched as a city truck weighed in at the entrance to the station, trundled up the ramp, and tipped its compacted load of paper. Freed from its confines, the clot seemed to hesitate on the brink of the hopper, as if reluctant to make the forty-five-foot plunge. Then slowly it began to tumble: a white-and-brown cascade of cardboard, loose paper, and junk mail that gradually picked up speed. The ten-ton mass hit the barge hard. There was a bounce and a whooshing sound, and then a tremendous updraft pulled a portion of the paper back toward the truck. The lightest stuff rose ten, fifteen, twenty feet. It reached its zenith, the force of gravity took over, and the paper wafted back onto the barge. Within hours, it would be transformed into a liquid, and then the mass with a million edges would be transformed once again into a single continuous ribbon.

The decision to recycle paper seemed like a no-brainer in New York. The stuff was picked up in its own container at the curb, transported to a local plant by barge, and turned into a product for which there was a fairly strong market. The city even made money off the deal. Not surprisingly, it was hard to find anyone, in government, in the advocacy community, or in the recycled-paper business, who would complain about the system. And yet paper and paperboard still made up nearly 40 percent of the garbage that

New Yorkers—and millions of other Americans who had curbside recycling programs in their cities—sent to landfills. Even in the best of times, New York's Department of Sanitation diverted just 19 percent of the city's residential paper from landfills or incinerators. It was worrisome that something that seemed so uncontroversial and took such minimal effort on the part of residents was so pitifully attended to. If we couldn't do right by paper, I wondered, how were we going to manage with materials that had a more complicated recycling profile?

Clearly, doing right by paper isn't a black-and-white issue to big corporate buyers. When I asked Little, Brown about printing *Garbage Land* on recycled paper, I was told that my book would be printed on paper that uses approximately 50 percent less virgin fiber than other book papers, that it would be produced by a manufacturing process that uses fewer bleaching chemicals than other papers, and that more than 80 percent of the fiber would come from lands certified by the Canadian Standards Association as sustainably managed. Sounds pretty good, right? But in fact, this statement says nothing about recycled content. I was happy to learn that the virgin fiber came from CSA-certified forests, but I'd be even more pleased if those forests were certified by the Forest Stewardship Council, which is more widely endorsed by conservation groups. And it doesn't get to the heart of the problem: not using new trees at all.

In the case of this book, there wasn't much I could do after suggesting a reputable company selling well-priced, high-quality recycled stock. Little, Brown was owned by Time Warner, and because Time Warner buys vast quantities of paper for all its books and magazines, it receives a deep discount. And then there's this: Like any author, I'm hoping many people will buy my book, which will result in even more paper usage. I'm still hoping Time Warner will eventually take the leap into all-recycled terrain, and in doing so radically alter the paper landscape, just as California lawmakers affect the automobile market when they introduce new emissions standards. Perhaps future editions of *Garbage Land* will be printed on recycled stock (or mostly sold electronically). Given the way we're decimating our forests, there may not be a choice.

Chapter Seven

Hammer of the Gods

In 1928, a young metals trader named Hugo Neu left his native Germany for New York City. By 1945, he had started his own scrap business, buying and selling metals all over the world. Most of his stuff came from peddlers and auto wreckers, but he didn't shy away from shiploads of bombs, which he defused, dismantled, and sold. It was a great time to be in the scrap metal business: the government was urging citizens to contribute to the war effort by scouring their attics, basements, and garages for any bit of salvageable metal.

In the 1950s, Neu tried to get his hands on New York's household metal, which was commingled with residential garbage. But the city had no interest in reviving a sorted waste stream, and the Neu Corporation carried on with its bulk metal business. The invention, in the sixties, of cutting blades that could make mincemeat of the thickest steel allowed Neu to start shredding metal. Smaller pieces were easier to transport and to melt into new shapes: business expanded exponentially. Today, the Hugo Neu

Corporation runs the fifth-largest steel recycling operation in the United States (out of about twelve hundred dealers). The company has scrap yards in New England, Los Angeles, Hawaii, New Jersey, the Bronx, and Queens. It was to this last yard that my household metal—less than four pounds of it a month—now went.

I stopped in at the Hugo Neu Corporation's Manhattan headquarters on a winter morning to see Wendy Neu, the company's vice president for environmental and public affairs, who also happened to be the wife of John Neu, the late Hugo's son. We'd met a few months earlier at a recycling roundtable. She was tall and exotic looking, with long black hair and olive skin. Her boots were high heeled and her suit, for an industrialist, incongruously hip. When Wendy agreed to meet with me, my pulse quickened: I had a date with the coolest chick in the room.

The Hugo Neu offices were sleek and muted, with arty photographs and a lot of big plants. They seemed more suited to an independent film company than to scrap metal dealers. I suspected this was Wendy's doing. In ways subtle and bold, she was trying to change the face of her industry.

"Recycling is still a niche business," Wendy said from her cluttered desk as she slowly peeled a clementine. "There's a junkyard, scrap yard mentality—you set up on an empty urban corner. There's no OSHA [Occupational Safety and Health Administration] standards, no environmental compliance." As the head of the government relations committee of the Institute of Scrap Recycling Industries (ISRI), Wendy was trying to raise the environmental bar. "We're dealing with storm water runoff, with oil-based materials, mercury from car switches, appliances, anything. We've got PCBs, Freon, paints. We recycle ten-year-old cars, which contain stuff that's since been outlawed."

Her scrap yards, she said breezily, received Notices of Violation from the state "all the time." It was her idea to contact the local environmental groups—the wetlands scientists, the bay keepers—to find cleaner ways of doing business. "The idea in the past was for scrap companies to keep hidden. We made a policy decision to let people in. It made good business sense, forget about the reality of cleaner workplaces. And we want small companies to toe the line, too, because we need a level playing field. We're all competing for business, but we all have to put in the same capital investment."

Because I was still having trouble connecting scrap yard images with Wendy's green leanings, I asked how she'd come to work here. After graduating from American University, she said, she'd started studying for a master's in social work at Rutgers. She worked at Trenton State, a maximum-security prison. "But I was bad at it. I got too close to the inmates. Basically, I recycled them—they'd go out and then come back." After that she worked at a youth correctional facility in Yardville, New Jersey. "It was sex offenders and murderers—the trash no one wants," she said. She smiled ruefully. "I've been consistent in that regard."

In her twenties and unemployed, she came to work for Hugo Neu. Her father, a marine surveyor for Lloyd's of London, got her the job: the scrap metal company was one of his clients. John Neu had advised Wendy against taking the job. "He said, 'There are no women here; you'll have to deal with Greek sea captains, with foul language.' I said to him, 'I've been with working with sex offenders and murderers in group therapy, working with death row inmates.'"

Wendy laughed and popped a crescent of clementine into her mouth. "I started in the traffic department, writing letters of credit. I was the first woman here who wasn't secretarial. Then Hugo sent me on the road, to look at equipment. When he got sick, I worked for another gentleman, selling nonferrous metals all across the country. After a couple of years I thought it was time to do something else, but John took me out to dinner and asked if I'd stick around. I didn't even like him at first. I liked his father. But we got involved, and he talked me out of quitting."

After Wendy and John married, she enrolled in law school: she wanted to work in prisons as a public defender. "I decided after three months of school that I'd make a lousy lawyer," she said, "but I stuck it out." The prisons never got Wendy back, but the Hugo Neu Corporation did.

* * *

Of all the materials recycled in this country, metals have the longest history of being collected and refashioned into new goods. The market for scrap has always been strong: Alan Greenspan is said to look at its price as a leading economic indicator. Recycling scrap is cost-effective, and working with clean recycled steel in electric arc furnaces, instead of virgin ore in blast furnaces, requires a third as much energy (because it avoids mining, processing, transporting, and converting ore to iron), cuts air pollution by more than 85 percent (because all that digging, transporting, and transforming produces large amounts of greenhouse gases), and cuts water usage by 40 percent (because iron ore isn't being cleaned and cooled).

According to the Steel Recycling Institute, which is headquartered in Pittsburgh and likes to frame things in terms of local football, the steel saved by recycling more than forty-six million appliances (the number recycled in 2003) would build 189 stadiums the size of the Steelers'. More than two appliances were recycled for every NFL fan who attended a regular season game in 2002 (put in less convoluted terms, 85 percent of appliances got recycled). All very good news, because every ton of steel that is recycled saves 2,500 pounds of iron ore, 1,400 pounds of coal, and 120 pounds of limestone—the amount of materials it would take to mine and refine new steel. Over the course of one year, the steel recycling industry conserves enough energy to power about eighteen million homes (some of which, no doubt, contain Steelers fans) for twelve months. All that, and recycling reduces landfilling and incineration, too.

Recycling aluminum also generates huge savings: it takes five million tons of bauxite ore, and the energy equivalent of thirty-two million barrels of crude oil, to produce a million tons of beer or soda cans. Make new cans from the old ones, however, and all that bauxite, plus significant amounts of petroleum coke, soda ash, pitch, and lime, stays in the ground. By averting the transformation of these materials into new aluminum, recycling cuts energy use by more than 94 percent and avoids the same amount of air pollution.

Making new metal out of old metal sounds green and right

thinking. But the mental image conjured by Earth Day, of cheerful volunteers plucking tin cans from fields of wildflowers, has little to do with the reality of a scrap yard, as I realized on a Friday morning in May. It was a textbook day for visiting a metals handling facility—not too cold, not too hot, a little breezy, no rain. But still it was obvious: processing old metal was a dirty job—labor intensive, highly polluting, noisy, dusty, and ugly.

Steve Shinn greeted me in his office at Hugo Neu's Jersey City scrap yard, which was set alongside Claremont Channel, a brown rectangle of water that led out to Upper New York Bay. Shinn, the plant manager, was an athletic-looking six-footer in a short-sleeved plaid shirt and work pants. While I examined a series of strange-looking lumps on his desk, a mini-museum of ore, Shinn said, "Scrap is the first feedstock of a steel mill, not virgin. Scrap is cheaper, and the industry has evolved to supply the mills with enough of it, at the right quality and the right price."

In 2003, about seventy million tons of steel scrap were recycled in this country, of which five million tons were exported, from six facilities, by Hugo Neu. The company's mandate was simple: it bought steel and shredded it; it bought shredded steel and shipped it; and it bought steel that it didn't shred but only cut into pieces and shipped.

I picked up a five-pound "knuckle" of iron, a rounded blob in a dullish silvery gray. "That's probably from a car," Shinn said. I turned over a lump of HBI, or hot briquetted iron, which weighed less than the knuckle, and a lump of HMS, or heavy melting steel. "We don't shred that, we cut it. It's made from pieces of machinery, metal stairwells, that kind of thing." I assessed a chunk of copper, from cooling pipes, and a chunk of brass, from decorative work. They were similar in color, but the brass was a lot heavier.

Next came a rough chunk of slag that had been scraped from the crucible of a steel mill's electric arc furnace, then a few ounces of shaley, shiny-looking aluminum. Shinn's phone rang, and my eyes wandered around the office, settling on a large map of the eastern United States on which every city with either a steel mill or a shredder had been marked with an arrow. On another wall was a panoramic photograph, matted and framed, of Claremont Terminal: it depicted a loaded barge, warehouses, preening cranes, conveyor belts, and a cyclone, a tower that sucks dust from the shredding system. The tones were pink, gray, and brown, and the water was the same murky color as the air.

Shinn's scrap yard, which covered about twenty-five acres, buzzed, hummed, and beeped with the comings and goings of trucks, trains on a rail spur, 'dozers, loaders, and lifters. Steel dust and road dust and diesel dust floated on the breeze; my front teeth, after a few minutes, tasted of metal. Shouting at each other to be heard, Shinn and I threaded our way through a line of eighteen-wheelers queuing for the scales. All the surfaces here, paved or not, were muddy: a spray truck circulated several times a day, wetting everything down. My first impression of the yard was that it consisted of random heaps of stuff connected by dirt roads and conveyor belts. That impression changed little over the course of my tour. "We make big piles, then we move the big piles to make more piles," Shinn said happily.

We approached a fifteen-foot-high mound of fuzzy reddish clumps. "We call those meatballs," Shinn said. (Like a short-order kitchen, the scrap business devised colorful names for standard offerings: copper wire was "barley," yellow brass was "honey," and number 1 copper tubing "candy.") The copper meatballs were armatures from electric motors, possibly from washing machines, and were entangled with spaghetti lengths of BX armored cable. Hard by the meatball stack was a pile composed of stainless steel—sink fixtures, gears, bolts—that looked dull and gray. It had all been handpicked from a conveyor belt.

"We want zero waste," Shinn said. Indeed, his people searched desperately for every last bit of value in the endless stream of discards, much as the city's rag- and bone pickers had a century earlier, and the highly organized dump scavengers in developing nations did today. Anything that didn't have value to Shinn was a liability: it cost him money to dispose of things he couldn't sell. Until recently, employees at the scrap yard were required to scrounge for coins that fell from cars headed into the shredder. The effort netted the company thirty thousand dollars a year, but management decided the amount didn't justify the worker hours. (Now one of the biggest quality-of-life issues at the Jersey plant was the soda machine, which was messed up because employees fed it quarters that they picked up in the yard, coins that had barely made it through the six-thousand-horsepower shredder and its violently vibrating screens.)

I'd be hearing more about zero waste in the months to come, but not from the perspective of industrialists who abjure waste on economic grounds. (That is, unless they are industrialists who tolerate waste as a negligible expense or because they can pass those costs along to someone else.) Rather, I'd be hearing about Zero Waste, capitalized, from recycling activists who envision a future in which waste has been designed out of systems, in which nothing is buried or burned. Zero waste at a factory, mill, or scrap yard saves money; Zero Waste as a philosophy saves the earth.

Shinn and I moved on to a thirty-foot-high pile of HMS. "This is ISRI specification size 201 or 202," Shinn reported, between three and five feet long. In the jumble I spotted car wheels, motor parts, gears, metal gates. There was a tramp amount of junk in here, too—paper, aluminum, wood—that would have to be picked out and, yes, recycled. Shinn regularly bought big stuff that he cut into 201 or 202, like the rusted-out grain hopper that sat off to the side. If I'd noted the hopper in a farm field, I'd have considered it a blight on the landscape, symbolic of death or failure. Here, though, big derelict equipment was just marking time until it could be sold, shipped, and reshaped into something shiny and new.

"What's that smell?" I asked Shinn, sniffing.

"We've got a contract with a fragrance manufacturer. We're shredding perfume drums." A whole quadrant of the yard smelled like *Glamour* magazine. I took a step backward and blinked up at a yellow monster rising three stories over my head. Its caterpillar treads were seven feet high, and it had a sixty-foot arm and a grapple claw that looked capable of lifting a small house. "That's a material handler," Shinn shouted proudly over the din of trucks and an insistent clanging. "I converted it from a hydraulic excavator." I'd gazed down upon similar goliaths at the paper plant and the incinerator, but this was my first time getting a scrap's-eye view of such a machine. It made me feel soft-shelled and puny. The grapple could hoist six tons at a time, and it weighed 225,000 pounds. Next to it sat an excavator with a fifty-five-foot arm that ended in a magnetic fist, a disc six feet across.

We watched a front-end loader shove a pile of "white goods"—hot-water heaters and washing machines—into a tidy hill. A second grapple unloaded tire-free automobiles from the car carrier. Presquashed at other junkyards, they all had the low profile of convertibles. "The cars we get are about eight to ten years old," Shinn said. "They provide about half of our feedstock." Like a cat toying with a mouse, the grapple dangled its prey delicately, aligning each car with the conveyor belt that would move it steadily upward into the furiously beating heart of the operation, the steaming, clanking Prolerizer.

"What the hell is a Prolerizer?" the radio program *Car Talk* once asked. "You remember that wood chipper from *Fargo*? Take that, and make it about fifteen hundred times bigger. Add hammers and teeth. And start the process with a conveyor belt and a seventy-five-foot plunge. Beginning to get the picture? That's a Prolerizer."

Invented by Hymie, Sammy, Izzie, and Jackie Proler—brothers who ran a Houston-based scrap company and formed a joint venture with Hugo Neu in 1962—the Prolerizer has a six-thousand-horsepower synchronous motor and enormous blades that can convert whole cars to fist-sized chunks of scrap in thirty to sixty seconds. "There are six of these machines in the world, and Hugo Neu operates four of them," Shinn said as we stood fifty or so feet from the shredder, which from my view was just a red metal shed at the top of the eight-foot-wide conveyor belt. The cars rose at a forty-five-degree angle and then disappeared behind a rubber curtain. Billowing steam obscured the entrance, lending the scene an ominous Mordor feeling. Behind the curtain, Shinn said, cars plummeted onto the shredder's spinning rotor, which bristled with thirty-two bow tie–shaped blades that weigh three hundred pounds each. Shinn called the blades hammers. They were thirty inches long, and though made of a steel-manganese alloy, they lasted a mere twenty-four hours, such was the ferocity of their labors.

The Prolerizer had stood the company well, and Shinn spoke of the machine with some warmth. But soon he'd be saying good-bye. "We're getting two MegaShredders in here," he said. "Their hammers weigh twelve hundred pounds, and they can handle two and a half times the Prolerizer's capacity on an hourly basis." I felt a pang for the old shredder: the future was passing her by.

"What will you do with this one?" I asked.

"Sell it to a competitor far, far away."

Less than a quarter mile to the north, across the bulkheaded walls of Claremont Channel, a row of pastel-colored condominiums, originally offered with individual yacht berths, lined the waterfront. They were expensive homes, with panoramic views of Manhattan, landscaped grounds, and ferry service to the financial district. On the Hugo Neu side of the channel, a fifty-five-foot-high sound barrier made of vertically hung conveyor belts blocked the condos' view of the Prolerizer, but that's all. Swells who lived on the south side of the complex had an unimpeded view of everything else—the mountains of scrap, the trucks, sheds, and machines. It was an unapologetically, fully functioning industrial landscape.

"Do you hear from them much?" I asked Shinn, cocking my head toward the condos.

"All the time," he said. "But mostly when there's an explosion." Explosions were not uncommon. Sometimes auto processors incompletely drained gas tanks, and sometimes a discarded propane tank snuck through the inspection process. "It's just a bang," Shinn said. "It's a nuisance, but it's not a big deal." Water sprayed continuously inside the shredder, to cool the blades, to control dust, and to extinguish any incipient conflagrations.

"So there are no flames or anything?" I asked.

"Yeah," Shinn said, shrugging. "There are flames."

We poked into the mill motor room, adjacent to the shredder. In the dim light, I made out the Prolerizer's enormous flywheel. Upstairs, a computer monitor was built into a shabby wall, anachronistically modern in a greasy junkyard shack. The computer said the shredder's rotor speed was 400 rpm. "Kinda low," Shinn said, sounding annoyed. Actually, the speed was high, which meant the shredder's productivity was low: nothing was in its maw. We went up a different flight of metal steps, behind the shredder. It felt dangerous to approach this barely contained beast that made mincemeat of metal, but Shinn didn't blink. Imagining shards of steel zinging through the air and flames leaping from fumaroles, I slid my safety glasses back up to the bridge of my nose. Apparently no one worked close to the shredder, except for a grease-smeared employee picking meatballs from a conveyor belt thirty feet downstream. In his scrap-patched cubicle, he seemed oblivious to his proximity to apocalypse.

It was hard to talk this close to the Prolerizer. Every time a car hit the blades there was a loud *thunk,* which we felt through our boots. The motor roared constantly, overlaid with the pleasant *tink-tink* of falling glass and the clatter of much-abused metal. Shinn pointed into the billowing steam and shouted into my ear: a grate was positioned below the blades, he said, and under that was a vibrating pan. The grate was made of cast iron, eight inches thick. Over the course of a month, it withered away to just four. He showed me an old grate lying on the ground, but the rest of the Prolerizer's guts I had to imagine—the rotating magnetized drum that pulled steel from the first cut, the shakers and screens that drew off debris. From here on, everything moved over conveyors, which rose and fell in ramps all around us. They separated and came together in Ys, trundling their loads to magnets and eddy currents that dropped material onto other belts, at right angles and oblique angles, that brought them somewhere else. Always, piles were forming.

Leaving the Prolerizer, we came to a monadnock of shredded ferrous scrap. "Wow," I said, impressed.

"You think this is big?" Shinn said. "This isn't big. It's about thirty thousand tons." Over the course of a month, the Jersey yard bought and processed about 125,000 tons of metal, or one and a half million tons a year. "It sounds like a lot, but it's not a whole elephant," Shinn said. "It's spoonfuls of an elephant."

The scrap pieces in the mound of the moment were roughly the size of a football, twisted and torn and thoroughly three-dimensional. Naively, I had imagined shredded metal as generally even as Kellogg's cornflakes. Here I saw whole rotors, cans, a crankshaft, and a hubcap that had made it through the Prolerizer more or less intact. All the stuff in this pile was here for the same reasons: it was broken, it had failed, it had become obsolete, someone was tired of it, no one could find another use for it. The pile was the very definition of garbage. I especially liked the democracy of the heap: everything here, no matter where it came from or what purpose it had served, was going to end up in a crucible, melted and reformed into railroad tracks, reinforcing beams, wastepaper baskets, gardening sheds, and the roofs of Ford Focuses and Rolls-Royces that would one day be carted back to a scrap lot.

In the Jersey City purgatory between their old life and their new, cars presented all manner of headaches for Shinn. There were the explosions, of course, which could damage the Prolerizer. And there was mercury that leaked from headlamp switches onto the ground, along with brake and transmission fluid that escaped collection, and then seeped into the water. And then there was solid waste. Here and there on the lot towered thirty-foot-high piles of shredded automobile residue that had rained down from the Prolerizer onto a belt: glass and plastic and rubber and foam and plain old dirt. The pile was dirt colored and dirt textured: it looked uncannily similar to finished compost.

According to the EPA, the United States scraps ten million cars a year, leaving auto shredders with approximately five million tons of residue, or "fluff." Shinn bent down to grab a handful of the stuff. "See these tiny wires? That's copper. We're trying to find a way to get this out. It's worth fifteen cents a pound if it's insulated, twice that if it's uninsulated." Again, like any businessperson, Shinn didn't want to discard—or pay to discard—anything with potential value. (A year after my visit to Jersey City, I learned that Hugo Neu was trying to site a 130-acre, 350-foot-high auto fluff landfill in Navassa, North Carolina. When citizens of the community, which is mostly poor and black, learned that auto fluff contains mercury, lead, cadmium, chromium, arsenic, polyvinyl chloride, and PCBs, they organized to fight the fill, which had already been rejected by another economically depressed community, in South Carolina.)

Leaving the automobile residue behind, we walked toward the channel, where towering blue cranes waited to load fifty-thousand-ton ships that were bound, more often than not, for Asia. In 2002, Hugo Neu exported about 80 percent of its scrap. (The company was, in fact, the largest scrap exporter in the United States.) According to industry figures, exports of scrap iron and steel reached 11.9 million tons in 2003, almost double the tonnage of 2000. China was the largest importer, accounting for 3.5 million tons—about 30 percent—of the exports. Meanwhile, domestic use of scrap iron and steel has been holding steady at about seventy million tons a year.

In 2003, China became the first country ever to import more than $1 billion of American scrap, according to the newspaper *American Metal Market*. Chinese fabricators offered more money for scrap metal than could American companies; Chinese laborers wielding hand tools and working for cheap would disassemble the jumbled chunks, shape the metals into valves and faucets, then sell them back to the United States at prices American labor couldn't beat. In response, steel industry trade groups were asking the government to temporarily limit scrap exports.

Shinn filled two or three ships a month, sometimes with scrap and sometimes with five-foot lengths of steel cut by guillotinelike shears. "The beams from the World Trade Center were too thick even for the shears," Shinn said. "We had to cut those with a torch."

To reach Jersey City this morning, I had driven south down the New Jersey Turnpike. Traffic was light, and I had time to contemplate the absence of the twin towers on the lower Manhattan skyline. Hugo Neu had won the contract with the city to recycle the towers' steel. Working around the clock, workers had hauled beams straight from the Fresh Kills landfill and into Claremont Channel. The company also sent its barges directly to Ground Zero. Those chunks of copper and brass that I'd plucked from Shinn's desk? They had once been part of the towers.

"We had to handle the steel very, very quickly," said Wendy Neu, when we'd spoken in her office. "It was a gargantuan effort; we had never handled anything like that. We had to lease equipment and hire crews. Trucks were bringing ten thousand tons a day. We were operating twenty-four hours a day, and we had to keep selling it, to make room." Ultimately, the company bought 225,000 tons of steel—about two-thirds of the total—probably for between seventy-five and a hundred dollars a ton. Cut and shipped to Asia, the steel would easily have fetched at least twice as much. Asked about the company's profit, Shinn would only say, "We didn't make a lot; our profit came from the volume." (After the World Trade Center material was sold, the price of scrap steel started to rise, reaching more than three hundred dollars a ton by 2004, according to the Emergency Steel Scrap Coalition.)

Admittedly, the WTC wreckage had far more emotional resonance than your average load of steel, and yes, the Hugo Neu Corporation had efficiently met a need, removing debris quickly so the city could recover. But there was nothing heroic about the company's efforts. Hugo Neu had, in the end, done what it always did. It profited by trading waste.

Driving home, I realized that I hadn't seen any tuna cans in the Claremont yard. I called Shinn, and he assured me they were there. Some of them were surely mine. Every Wednesday morning at about 7:00 a.m., a packer truck from the Brooklyn South 6 garage grunted and hissed up my street. Usually it was James Sheehan and Dominick Basso on truck CN210. One of them drove while the other stepped off the back and quickly tipped my building's plastic barrel, which was a quarter filled with metal cans, into the hopper. (When glass and plastic recycling returned, in two years, our fifty-gallon barrel would, each week, be full, and thirty pounds heavier.) Sheehan and Basso switched every few blocks. The work seemed especially monotonous, if not particularly strenuous. No brownstone's metal-recycling bin weighed more than a few pounds.

"It's a lot of work," Sheehan said to me one day. "The route is an entire section now."

"How long is that?" I asked. Sheehan gave me the san man's salute, a two-shouldered shrug. (There was another salute, a more Seinfeldian movement in which the right hand was raised and then flipped downward in a "fuggedaboutit" gesture. You'd use it, for example, in a mongo situation. "You want this VCR?" Hand gesture: "Take it! I got one a-reddy!") Without plastic and glass in the recycling bins, the guys on metal could go for eight hours and never pack out. The next shift took the same truck, and when it held four or five tons—after two or three days—it was relayed to the Hugo Neu scrap yard in Long Island City, in Queens.

There, just as in Jersey City, a grapple would drop my tuna can onto the conveyor belt, and the can would dive into the whirring Prolerizer, then fall onto a pan and then onto another conveyor belt, which would send it past the pickers, who would ignore it. The can would drop down to recombine with another stream, travel up a stacking conveyor, and plink down on a pile of shredded scrap. When the pile got big enough, it would be barged to Jersey City and loaded onto a ship, along with shredded cars and washing machines, bound for Asia.

I didn't buy soda, and if I did I'd surely redeem my aluminum cans for nickels, much to the Neus' dismay. Nationwide, beer and soda cans are the most-recycled consumer product. But their rate of return fell from a peak of 65 percent in 1992 to a twenty-three-year low of 44 percent in 2003, when 820,000 tons of aluminum cans were trashed. The Container Recycling Institute estimates that more than a trillion aluminum cans have been buried in landfills since 1972, when industry started keeping records. The amount is nearly equal in weight to the world's entire annual output of primary aluminum ingot. If all those cans were dug up, according to the Institute's Jenny Gitlitz, they'd have a value of $21 billion at today's scrap prices. The entombed cans have raised the issue of landfill mining, which might become common practice as natural resources disappear. Until then, waste managers will bury cans, and aluminum manufacturers will fight bottle bills while busily digging new bauxite mines and refineries, and adding smelter capacity in Iceland, Mozambique, and Brazil.

The material that Hugo Neu bought from New York's Department of Sanitation—about 6,200 tons a month of bulk and household metal—amounted to barely 5 percent of the company's business, a mere 74,400 tons a year. It was nothing, really. And yet the ratio echoed history: by the beginning of the twentieth century, when the scrap metal business was already very well organized, dealers welcomed all manner of small household goods, but the bulk of their trade consisted of the same feedstock it did today: rails, train cars, machine parts, beams, pipes, and tanks.

My personal contribution to Hugo Neu was tiny, but my efforts, along with those of other city residents, brought $2.2 million a year to city coffers. The low numbers of cat food cans and cookie sheets didn't discourage John Neu: he was convinced he could get more metal out of households. "We need better education," he told me. "We'll do it ourselves, and we'll try to get some nonprofits involved as well." He spoke highly of San Francisco's much-vaunted recycling program. I reminded him that San Francisco, which Wendy Neu and I had heard so much about at the recycling roundtable, and which I would soon visit, charged citizens to throw away garbage, while recycling was absolutely free.

Before I left Claremont Terminal, I had asked Steve Shinn if he wanted the metal tops from my peanut butter jars. "Yes, but not if they have a plastic coating." I assured him my brand did not. Tuna and pet-food cans have a plastic liner to protect food, he said, and the plastic could, in large volume, alter the chemistry of the melted steel. But, again, Hugo Neu wasn't handling a lot of cans as a percentage of the whole, so it was okay.

"Do I need to take paper labels off cans?"

"No. It's pretty hot inside the Prolerizer. It takes the paint off, too."

"How about my beer tops, do you want them?"

Shinn considered for a moment, then said, "It would go through the shredder. It would follow the process and probably end up in the suck-and-blow system, in the cyclone. A beer cap is light. It would be pulled into the vacuum system, to the stream of raw, nonferrous debris. From there it would be separated from the nonferrous material, picked up by a different magnet, and returned to the steel scrap pile."

"That sounds fairly circuitous," I said.

"Yep," Shinn said, not sounding daunted in the least. He was, after all, a man who sought value in tiny copper wires, in pennies that fell from heaven.

Chapter Eight

Mercury Rising

For some months now, I had been diverting dozens of pounds of bottles, cans, paper, and kitchen scraps from the landfill in Bethlehem, Pennsylvania, though the folks at IESI probably had not noticed. Now I turned my attention to household hazardous waste, a category that, were IESI following the letter of the law, they'd surely appreciate my efforts to divert from their tipping floors, trucks, and landfills. "Hazardous waste" conjured leaking casks of picric acid and loose bundles of TNT, neither of which I had on hand, but once I started looking around my house, I came upon all kinds of noxious materials. There was rubber cement and superglue in my junk drawer (they contain a handful of chemicals listed by the EPA as hazardous), an ionizing smoke detector on my hallway ceiling (it was made with a small amount of americium-241, which has a radioactive half-life of 458 years), a broken thermometer in the bathroom cabinet (mercury is a potent neurotoxin), paint thinner in the basement (turpentine and mineral spirits are both flammable and toxic), a bottle of windshield wiper solution in my closet (it contains methanol, which damages the nervous system, liver, and kidneys), and rechargeable batteries all over the place (they contain heavy metals, like cadmium and nickel, plus flame retardants that outgas poisons). None of this stuff was supposed to go into the regular garbage, but I had a sense that very few people knew or cared about this detail. San men aren't bloodhounds (except when it came to fine jewelry or bodies wrapped in rugs), and most of those items are small and easy to hide.

In the fictional world of a trash-conscious New York, residents would, with a spring in their step, collect this hazardous waste and deliver it to designated drop-off sites within their boroughs. When I telephoned the sanitation garage near my drop-off site (which didn't have its own phone number, and only the vaguest of addresses) and asked what they did with old paint, a san man suggested I just stick something absorbent in my cans and send them out with the regular trash. I had a sneaking suspicion he was offering me, over the phone, the official san man's salute.

"And what about the other hazardous stuff?" I asked. "What do you do with that?"

"We recycle it."

"How?"

"You'd better call public information," he said. "Whatever information is public you can find out." Oh, how I wished. Either the city didn't know what happened to the solvents and thermometers and lead batteries (banned by landfills in twenty-nine states, including mine) dropped off in the boroughs or it didn't want to say. For weeks my messages went unanswered, and I started to wonder if my efforts to keep waste from the municipal stream were worth it. Was I diverting stuff from place A only to have it return, after a circuitous journey, to place A?

I started calling stores that sold rechargeable batteries, the little black packs that make our cordless shavers, drills, camcorders, cell phones, and remote-control toys go. Retailers that partnered with the Rechargeable Battery Recycling Corporation (stores like RadioShack, Staples, Wal-Mart, and Target) were supposed to collect this stuff and ship it to a metals reclamation company. But I suspected, because no one gave me a straight answer, that harried clerks often slipped these small nuisances into the trash. Still, despite my suspicions, a fair percentage of batteries did make it back into the system. Over the past decade, the RBCR claims to have kept more than twenty million pounds of nickel cadmium, nickel metal hydride, lithium ion, and small sealed lead batteries from landfills and incinerators. Where did they go instead? To a factory thirty-five miles northwest of Pittsburgh.

Inmetco was a quiet and unassuming place, at least according to its manager of environment, health, and safety, John Onuska. "We don't have any dead geese hanging on the fence or toxic lakes," he offered. Inside the plant, battery packs were smashed apart, to remove the plastic from the battery cells, which were then dried out and heated in a crucible. Cadmium was vaporized, condensed to a liquid, and chilled to form flattened pellets, called shot. Inmetco sold the shot to battery manufacturers and to fine-glass makers, who used it to make bright red and yellow pigments. The nickel and steel that remained in the crucible were cooked up with steel dust and other mill scraps in a smelter to produce a "remelt alloy," which was sold to manufacturers of stainless steel.

Recycling metals, like much of the recycling industry, is a dirty business, the absence of dead geese notwithstanding. But it spares the earth the far larger insult of mining virgin metals, with all its attendant energy use and pollution. Extracting one ton of copper, for example, requires miners to move an additional nineteen tons of rock. According to a Commission on the European Communities report on battery recycling (the EU, having embraced the precautionary principle, is way ahead of the United States in formulating policies to avoid and recycle high-tech waste), working with recycled cadmium and nickel requires, respectively, 46 percent and 75 percent less primary energy compared with the extraction and refining of virgin metal. Using zinc recovered from alkaline batteries consumes 22.5 percent less energy than extracting it from primary resources.

Would I be breaking the law if a rechargeable battery slipped into my kitchen trash can? The EPA's Resource Conservation and Recovery Act regulates the hazardous constituents of batteries, but the law offers some latitude. Commercial enterprises that handle these metals are required to dispose of them in landfills designed for hazardous waste, but small businesses and households are exempt. The EPA cares less about alkaline batteries, since manufacturers agreed, in the late nineties, to phase out their use of mercury. (Still, the double-A's in your Walkman and the C's in your flashlight represent 80 percent of all batteries manufactured today, and they each contain trace amounts of mercury. Worldwide, manufacturers produce more than ten billion of these cells a year, creating hundreds of millions of pounds of solid waste.) "Are we supposed to take those batteries? No," Jim White, at the American Ref-Fuel plant in Newark, New Jersey, told me. "Might one possibly make its way past our inspectors? Yes."

In sixth grade, I attended a four-room schoolhouse with a wide-open basement. Mischievous children used to sneak downstairs at recess and comb through school supplies. Supervision was lax, and so after discovering a box of thermometers one rainy afternoon, we had nearly an hour in which to roll tiny balls of liberated quicksilver from hand to hand, bowl them across the floor, and lob them at one another's shirts. Eventually we got busted, but no one seemed concerned that the element with which we'd been playing so merrily was poisonous. Even the ancient Romans knew that mercury was bad news: they got convicted criminals to extract it from cinnabar ore, then they mixed it with gold to gild objects. These miners lasted an average of three years before they lost their hair, their teeth, and eventually their sanity.

Wildfires and volcanoes produce about a third of the mercury in today's atmosphere. Coal-fired power plants and incinerators that burn mercury-containing wastes (like household batteries and thermostats and computer circuit boards) generate the rest. (According to a study by Barr Engineering for the EPA, coal-burning electric utilities emitted an estimated forty-eight metric tons of mercury into the atmosphere in 2000.) Again, the problem with mercury is that once it becomes airborne, it mixes with rain and snow, then settles on lakes and waterways, where bacteria convert it to methylmercury, which works its way up the food chain. Over time, exposure to elemental mercury causes permanent, and sometimes fatal, kidney and neurological damage (plus, of course, hair and teeth loss).

Still, mercury abounds in consumer products—in fluorescent lamps, gauges, car light switches, and dental amalgams. In the United States alone, citizens innocently discard some seven million thermostats a year, each containing three to four grams of mercury. Thermometers contain a half-gram of mercury. That doesn't sound like much, but the element is extremely potent. One Sweet Tart–sized mercury battery is enough to put a six-ton load of garbage over the fed's allowable limit for solid waste. In landfills, mercury leaches into groundwater, where bacteria transform it to the evil methylmercury. According to a study widely quoted by the EPA, it takes the settling of just one gram of vaporized mercury upon a lake of twenty acres to unleash a yearlong fish consumption advisory on that body of water. As regular as fishing season's opening day, state agencies issue mercury warnings for thousands of lakes across the nation.

The industrial world bristles with entrepreneurs jostling to get their hands on mercury from household products and industry. Chlorine producers who have given up mercury in favor of new technologies, for example, have vast quantities on hand. But the hopeful quicksilver merchants aren't neutralizing neurotoxins: they are spinning a liability into a commodity bound for developing nations that haven't banned its use. China, for example, relies on mercury for gold mining. India also imports large quantities for use in medical instruments and other manufacturing processes. Some environmentalists claim that this overseas trade keeps mercury prices artificially low, which increases its use in places that lack environmental regulations. One solution is to halt the sale of mercury and stockpile the old stuff in repositories. The United States currently stores 4,890 tons of mercury (in steel flasks inside thirty-gallon steel drums) at four Department of Energy warehouses in New Jersey, Ohio, Indiana, and Tennessee. The Pentagon is agitating to consolidate it all in one location. Until then, shifting our mercury overseas keeps developing nations from mining virgin mercury. But while the commodity flows in one direction—from nations that enforce environmental protections to nations that do not—its hazards are multidirectional. The element vaporizes with ease, and so its poison drifts with the breeze. The local hazards of mercury are global.

One Saturday morning, I cleaned out thirty cans of paint and painting paraphernalia from my basement and set out to find, with my neighbor Tony and his fifteen pounds of spent batteries, Brooklyn's hazardous-waste drop-off site. It was supposed to be near Brooklyn's Sanitation Garage 11, in the Gravesend neighborhood, but when we finally found that garage, the san men enjoying an outdoor smoke told us the drop-off site hadn't been open for a year.

"What are we supposed to do with all this poisonous stuff?" Tony asked.

"Throw it in the trash," a burly guy suggested. While Tony fumed, I asked another worker where the drop-off site had been—I just wanted to see it—and he pointed toward a small trailer parked in an adjacent lot. "Good luck saving the earth!" trilled the first san man as we drove away.

After winding around parked snowplows and mounds of winter salt, we came to a chain-link fence. On one side was a crowd of angry Brooklynites; on the other, the trailer and a wispy-looking san man who claimed he couldn't open the gate. "I could squeeze through with my paint cans," I proposed.

"We're closed," he answered, then radioed his boss for help.

"The sign says you're open," said a young man in a Yankees cap. He had a liter of hydrofluoric acid in his trunk and he didn't feel like carting it back home.

The name of the acid should have rung alarm bells with the san man, but he kept a poker face. In 1996, a drum of this acid, which is used to etch glass and clean aluminum, exploded from under the packing blade of a garbage truck and showered san man Michael

Hanly, a twenty-two-year veteran of the force. He died, from inhaling the fumes, that day in the hospital.

When the supervisor appeared at the chain-link fence, the rain of abuse from a growing crowd of residents who'd been told this site was open was redirected toward him. "I'll take the latex paint and the batteries," the super said to me. He wouldn't take the Yankee fan's acid or a jug of windshield wiper fluid. "Call the DEP about them liquids," he said, easily sliding the "broken" gate open. (A Department of Environmental Protection employee later directed the Yankee fan, apologetically, to a private disposal company in Queens, which offered to pick up the acid for about three hundred dollars, or charge him fifty dollars if he'd drive it over himself. He chose the latter.)

Wade Salvage, based in Camden County, New Jersey, would eventually collect the latex paint I left at the drop-off site, along with a number of other household hazards. When I asked Andrew Wade, the company's president, what he did with these materials, he revealed as little as possible. "I've got twelve locations," he said. "I've got recycling and reprocessing customers." Could he be more specific? Only after Wade realized I wasn't "one of the weirdos" that continually hounded him did he disclose that batteries ended up at Inmetco (the recycling plant without dead geese, near Pittsburgh); that liquid mercury went to retorters who processed the element and sold it for reuse in various switches; and that fluorescent light bulbs went to a recycler who crushed the tubes and sent the glass to cement factories, the aluminum ends to metal scrappers, and the phosphorus powder to a retorter who extracted the mercury for resale and sold the phosphorous for filler. Paints, transmission fluid, and solvents were blended into an "alternative" fuel for paper and cement factories (it burns far hotter than coal or oil). Oil filters were banged into scrap metal. So-called poisonous waste, like insecticides, was tipped at incinerators permitted to accept household hazards.

Like any businessperson, Wade saved up his little piles of waste until he had piles big enough to sell on the open market or pay to have shipped and tipped. New Jersey didn't make it easy to store this stuff—the state's Department of Environmental Protection was partly funded with fines and fees—so my paint cans and batteries traversed the Garden State to land at Wade's "consolidation site" in Philadelphia. From the City of Brotherly Love, Wade told me with an air of great mystery, "it goes out to my various customers."

When Tony and I got back to Park Slope, I stacked eighteen cans of enamel paint, turpentine, and sealant in the bottom of my trash can. The san men, on pickup day, took everything piled on top but left the painting supplies. The clerk at the Brooklyn 6 garage, when I phoned to complain, advised me to open the cans, let the paint dry out, then put them on the curb for the next pickup. I set up my paint farm, but after a week nothing had solidified. Beginning to get annoyed, I started randomly asking san men for advice. Finally, I had a consensus. After checking to make sure that no one was looking, I banged the cans shut, hid them inside a tightly tied black plastic sack, and closed the lid on my pail.

Every day of the week except Sunday, New York's san men heave computers, cathode ray monitors, printers, cell phones, fax machines, and other electronic paraphernalia into the back of their packer trucks. The bigger components lie on the sidewalks before pickup, sometimes for days, and are incorporated into the street life of the neighborhood. Construction workers take cigarette breaks perched atop seventeen-inch monitors; a CPU becomes a plinth for an azalea. This stuff, when the garbage truck comes around, isn't mongo: it is junk, likely less than two years old (according to the alarums of environmental advocates) and already obsolete.

Across the nation, electronic waste is accumulating faster than anyone knows what to do with it—almost three times faster, in fact, than our overall municipal waste stream. According to the National Safety Council, nearly 250 million computers will become obsolete between 2004 and 2009. Carnegie Mellon University researchers have predicted that at least 150 million PCs will be buried in landfills by 2005, and by the following year, predicts the

Silicon Valley Toxics Coalition (SVTC), some 163,420 computers and televisions will become obsolete every *day*. Where will all these gizmos go, and what impact will they have when they get there?

Before I started to poke around my garbage, I had no clue that the computer sitting so innocently upon my desk, a virtual extension of my body, essential to my work and increasingly useful for buying more stuff, was such a riot of precious but pernicious materials. The average desktop monitor contains nearly four pounds of lead. (Electronic waste is the largest single source—about 40 percent—of lead in municipal dumps.) Printed circuit boards are dotted with antimony, silver, chromium, zinc, lead, tin, and copper. Cell phones have their own periodic arsenal: arsenic, antimony, beryllium, cadmium, copper, lead, nickel, and zinc. Exposure to these metals has been shown to cause abnormal brain development in children and nerve damage, endocrine disruption, and organ damage in adults.

Tapping away at my keyboard was probably doing me little harm, I figured, but it wouldn't take much for my sleek little ThinkPad to morph into a corrosive contaminant. Crushed in a landfill, it would leach metals into soil and water (remember, all landfills eventually leak); in an incinerator, it would exhale noxious fumes, including dioxins and furans, that would taint both fly and bottom ash. Everything must go somewhere—the environmental scientist Barry Commoner said it long ago, and I understood it implicitly now. But when I learned that fourteen of the fifteen largest Superfund sites were metal mines, it made me wonder: if we continued to take metals from the ground, solder them into consumer electronics, and then dump those components into fresh holes in the ground, wasn't it only a matter of time until those holes became Superfund sites, too?

I felt guilty about the afterlife of my computer, but the processes that gave birth to this miracle machine were equally troubling. According to a United Nations University study on the environmental impact of personal computers, it takes about 1.8 tons of raw materials to manufacture your average desktop PC and monitor. According to the EPA's Toxics Release Inventory, mining operations in 2003 released nearly 3 billion pounds, or 45 percent, of all toxics released by US industries: mining is the nation's largest industrial polluter. And we are a nation that *has* environmental laws. In the rush to supply our demand for new copper, coltan, gold, silver, and palladium—the stuff that fuels our 'lectronic lifestyles—African and Asian nations are tearing up their hillsides and hollows. Some gorilla populations in the Democratic Republic of Congo have been cut nearly in half as the forest has been cleared to mine coltan, a metallic ore comprising niobium and tantalum. Coltan is a vital component in cell phones, of which Americans discard about a hundred million a year.

Could a computer be recycled? I'd heard some murmurings on the subject, but computer recycling seemed to mean different things to different people. At one end of the spectrum were individuals giving or selling their working computers to others in need. At the other end were dealers who accumulated broken computers, cannibalized them for parts, and junked the rest. How did they get their source material? I had a chance to find out when my network router quit connecting me to the Ethernet. I relegated this mysterious black box, a chunk of plastic the size of a hardcover book, to my basement until a local recycling group organized an e-waste drop-off. Around the nation, charities, environmental groups, and municipalities were organizing similar drives, some of which collected nearly twenty tons in a single day.

I arrived at my collection site, at the north end of Prospect Park, to find several folding tables shaded by white tents and patrolled by clipboard-toting volunteers. The tables were laden with unwanted monitors, scanners, TVs, cell phones, keyboards, printers, mice, cables, and speakers, many of which had absolutely nothing wrong with them beyond a bit of dust and, in the case of the computers, a processing speed that only yesterday seemed dazzling. The spirit of bonhomie under the tents reminded me of the city Parks Department's Christmas-tree mulching parties, except that instead of returning nitrogen to the earth, we were, presumably, returning precious nuggets of metal to technology. (The parallel was made manifest at another city e-waste event, where volunteers offered fresh compost to anyone dropping off a stale computer.) Passersby pawed through the electronic casbah, taking what they wanted for free. The representative from Per Scholas, a Bronx computer recycler founded in 1995 to supply schools and other nonprofits with hand-me-down computers, could only look on stoically as the good stuff—which he could refurbish and sell—disappeared. The bad stuff—which included my router—was headed his way.

And so was I, on a drizzly winter afternoon. Debarking from the subway in the South Bronx, I made my way under one elevated expressway, across eight lanes of traffic, over the empty loading dock of a rehabbed brick factory building, up a freight elevator, and through a low defile of shrink-wrapped computer monitors stacked on wooden pallets. Ed Campbell, Per Scholas's director of recycling, led me into a large open room where teams of technicians wiped computer hard drives clean, then loaded the machines with Pentium II microprocessors, memory, and mice. The reconditioned computers, collected from corporations and institutions that paid Per Scholas ten dollars a machine, would be resold, at low cost, to "technology-deprived families." According to Campbell, Per Scholas's efforts kept some 200,000 tons of electronic waste from landfills and incinerators each year.

I watched a technician in training squirt a monitor with Formula 409, and then Campbell took me to see the darker side of the computer recycling revolution, where a cavalcade of monitors, some of the most powerful symbols of capitalism's success, were being smashed, one by one, to smithereens. The broken-down Dells, Apples, and Gateways—most of them collected at community drop-off events—trundled up a conveyor belt and into a shredding machine. "We call this a rough liberation," Campbell said. "Basically, we're crushing the computers into glass, plastic, and metals." Hidden inside the machine's bland carapace, a series of magnets, eddy currents, and trommel screens separated the shards and spat them into yard-high cardboard boxes called gaylords: ferrous metals here, nonferrous there, plastic on one side, glass on the other. A Per Scholas spokesperson said the metals went to Pascap, a Bronx company that resold them to smelters; the plastic went to a company that melted and pelletized it for resale. Disposing of the glass, which contained lead, presented Per Scholas with its biggest headache to date.

"Glass is a liability, not a commodity," Angel Feliciano, the company's vice president for recycling services, told me. "We save it up until we've got a truckload, then we pay $650 a ton to a smelter who'll haul it away." Lately, the glass had been landing on the loading docks of the Doe Run Company, in south central Missouri. The company separated lead from glass through a process of heating and reducing. It used the resulting silica as a fluxing agent in smelters that produced brand-new sixty-pound ingots of lead from raw ore. The lead liberated from the matrices of CRT glass, said Lou Magdits, Doe Run's raw-materials manager, was combined with lead recovered from car batteries, ammunition, and wheel weights. And where did all this recycled lead go? "Into car batteries, ammunition, wheel weights, and new CRTs," said Magdits. (In Peru, where Doe Run operates a lead, copper, and zinc plant, significant amounts of sulfuric acid rise into the air and fall down as acid rain. Farmers in La Oroya, home of the smelter, have charged Doe Run with contaminating their fields. In 1999, Peru's Ministry of Health determined that 99 percent of children in the area suffered from lead poisoning. The company, which bought the smelter from the Peruvian government in 1997, has entered into an agreement with the Health Ministry to reduce blood-lead levels in two thousand of the most affected children and claims that safety measures have decreased blood-lead levels in workers by 31 percent.)

One hundred percent of the material dropped off at my e-waste event was, to Per Scholas, "junk." But at least Per Scholas was handling its junk responsibly. According to the Silicon Valley Toxics Coalition, up to 80 percent of the material collected from well-meaning residents and businesses at e-waste events nationwide is bundled up and shipped overseas, mostly to China, India, and Pakistan. Perhaps half of those computers, the ones that function, are cleaned up and sold. But the remainder are smashed up by laborers who scratch for precious metals in pools of toxic muck.

Investigators from SVTC and the Basel Action Network (whose name refers to the 1992 Basel Convention, an international treaty that seeks to halt trade in toxic waste; the United States refuses to sign it) found men, women, and children in the Chinese village of Guiyu extracting copper yokes from monitors with chisels and hammers. Squatting on the ground, they liberated chips and tossed them into plastic buckets while acrid black smoke rose from burning piles of wire. After harvesting the easy stuff, the workers, who wore no protective gear, swirled a mixture of hydrochloric and nitric acid in open vats, trying to extract gold from components. Afterward, they dumped the computer carcasses and the black sludge in nearby fields and streams. Tests on the soil and water showed levels of lead, chromium, and barium hundreds of times higher than US and European environmental standards for risk. The accumulating carcinogens, here and in other Chinese coastal towns that accept e-waste, have contributed to high rates of birth defects, infant mortality, tuberculosis, blood diseases, and severe respiratory problems.

I was confounded by the way streams of postconsumer electronics divided and branched, with only a minute percentage of working computers returning to the headwaters of reuse. The computer recycling world, I said to Angel Feliciano, seemed suffused with weirdness. "Yes," he said to me slowly. "There *is* weirdness. There's inconsistency. Recycling, as it is today, is a great way for individuals to make money because they can pretend to be doing something." Feliciano was speaking, naturally, about the other guys, not his company. (In the months to come, though, I'd hear that Per Scholas had established a relationship with a New Jersey recycler that regularly bundles old equipment for sale overseas. At this point, Feliciano was no longer returning my calls.) "There's no regulation, and it's more profitable to do the wrong thing," Feliciano continued. "You ship it overseas. Or you have a sweatshop of people in this country working twelve hours a day for five dollars an hour, taking computers apart by hand and breathing this stuff in. You get into it for the short term. In five years, when you get out, the EPA will be taking a closer look."

It doesn't hurt the prospects of the unscrupulous that the general public is at once wildly enthusiastic to "recycle" its electronic waste and wildly ignorant about how this is done. I hate to say it, but it is a bit like the bad old days of curbside recycling in New York: folks had high hopes for the materials they set out, but so much of it—contaminated by food or bereft of an end market—ended up buried. Still, a peanut butter jar in the landfill is one thing: a circuit board in your drinking water is another.

Why is it so difficult to recycle computers righteously? For starters, it is dangerous, labor intensive, expensive, and unrewarding, in the sense that markets for the materials aren't always large or reliable. Then, there is a playing field tilted in favor of new production and the export of old. Some original computer manufacturers charge forty dollars for repair manuals on their products (instead of putting them on the Web) and lobby to make "gray market" refurbishing illegal in the developing nations where they sell their new models. At the state level, governments spend bond money to build incinerators and operate landfills, but most recycling centers have to balance the books on their own. The federal government encourages recycling and reuse, but it doesn't require it. "If we were paying what we should for virgin resources, e-waste recycling would be much more economical, and local governments perhaps could break even on e-waste recycling," said Inform's Eve Martinez, who has set up numerous e-waste collections. But mining companies, like logging companies and the oil and gas industry, continue to benefit from perverse subsidies. Under the 1872 Mining Act, corporations lease land at five dollars per acre, pay no royalties to the government on minerals they extract, and pass any environmental cleanup bills to taxpayers.

As the hazards of e-waste have worked their way into the news, some computer manufacturers have initiated take-back programs in which consumers wipe their hard drives clean, then swaddle, seal, label, and ship their large packages back to original manufacturers. The cost and the inconvenience discourage widespread participation by individuals, which is perhaps what those manufacturers had in mind. They don't want anyone buying a refurbished computer: they want consumers to buy a new one, preferably from them. IBM's take-back program was designed for the institutional user looking to either "recover value," in the company's words, or "dispose of obsolete assets" (in the first scenario, IBM buys back used equipment; in the second scenario, it hauls those ancient assets away, for a small fee). Hewlett-Packard's take-back program is friendly to individuals (the company even accepts computers and peripherals it didn't manufacture), but it is pricey. To mail my laptop, dead router, and one printer would cost me sixty-four dollars, minus the box and packing materials. (The company puts postage-paid labels and envelopes in some printer cartridge boxes.) When I asked staffers at one of the largest computer merchants in New York City about taking back my gently used IBM ThinkPad, they said they didn't do it, didn't know anything about it, and had never before been asked about it.

For its part, Massachusetts bans televisions and computers from landfills. Instead, it contracts with a company called ElectroniCycle, based in Gardner, Massachusetts, to process its e-waste. Harvesting material from drop-off events and retailers, ElectroniCycle recovers ten million pounds of electronics a year: technicians refurbish between 5 and 10 percent of their computers for resale; send another 5 to 10 percent to specialty repair houses; and smash the rest into fifty different categories of scrap, including plastic, copper, aluminum, barium glass, and leaded and mixed glass (which is recycled back into cathode-ray tubes). Reusable integrated circuits and memory cards are gleaned, then circuit boards are sent off site for recovery of gold, palladium, silver, and copper. Nothing goes overseas. In California, which also bans e-waste from landfills and restricts its shipment overseas, retailers that sell hazardous electronic equipment would soon be paying the state an "advance disposal fee" (collected from consumers) of between six and ten dollars per device to cover the cost of recycling. Still, manufacturers aren't required to participate in the collection or processing of waste.

Every time ElectroniCycle shunts a laptop to the reject pile, I imagine Michael Dell rubbing his hands in glee, like *The Simpsons'* Montgomery Burns when he finally fulfilled his lifelong ambition to blot out the sun. But computer manufacturers' freedom to pump out product without a thought for its afterlife is probably doomed. The Computer TakeBack Campaign, founded by more than a dozen social-justice and environmental groups, calls for manufacturers of anything with a circuit board to make "extended producer responsibility" (EPR) part of their credo. EPR would shift collection and recycling costs from taxpayers and government to the folks who make and promote these goods. Theoretically, the system would give companies an incentive to make computers and other gadgets that last longer, are made of reusable or fully recyclable materials, contain fewer toxics, and are swaddled in far less packaging.

In Europe, EPR is well under way. In 2003, the European Union adopted a directive that requires producers of electronics to take responsibility—financial and otherwise—for the recovery and recycling of e-waste. In Switzerland, which has far surpassed the goals of the EU directive, the cost of recycling is built in to the purchase price of new equipment. Consumers return e-waste to retailers, who pass it on to licensed recyclers. In the United States, almost half the states have active or pending e-waste take-back legislation: the proposals, naturally, cause electronics manufacturing trade groups to howl. Still, Maine—which has no tech industry to speak of—recently passed a law that requires manufacturers of computer monitors, video display devices, and televisions to finance a system for their environmentally responsible reuse and recycling. (Previously, the state had let residents stuff computers into their household trash if they could prove, through laboratory testing, that they were nonhazardous. Those who opted out of the expensive testing—and it's hard to imagine that anyone opted in—brought their e-waste to transfer stations, which paid recyclers to haul away and dismantle them.) Now Maine's computers will be delivered to consolidation points and sorted into mountains destined for their makers: Toshibas over here, HPs over there.

What those companies will do with them is, at this point, unclear. As Angel Feliciano said to me about the computer recycling world, there is weirdness.

What about other high-tech waste? Every month, according to the Worldwatch Institute, more than forty-five tons of CDs become outdated, useless, or unwanted. Responding to the glut, dozens of companies have sprung up to wipe clean compact discs, laser discs, and digital videodiscs. If the plastic discs can't be reused, they are shredded and blended into automobile parts, office equipment, and other products. (Meanwhile, the Disney Company is working in the opposite direction, selling DVDs that erase themselves after two days' exposure to air. Instead of renting a rewatchable disc, consumers buy something they use on Saturday night and slip into the trash on Monday.) Entrepreneurs degauss VHS tapes (and sell them for surveillance work) and collect and remanufacture ink and laser toner cartridges. (Other companies promote a green agenda but set the bar fairly low: Epson America collects ink cartridges from schools and other nonprofits, but instead of refilling them, a collection agency "converts [them] to energy through an environmentally sound incineration process at a licensed waste-to-energy recycling facility.")

Another set of entrepreneurs is scrambling to collect the tens of millions of cell phones that are stuffed into drawers when their owners switch services, or jilted when thinner models come along. Phones are refurbished, sold overseas, or programmed to dial 911 and donated to the elderly and to women's shelters. It is economies of scale that make these recycling efforts go, but the junked electronics in household waste streams are often too few and far between to make municipal collections worthwhile. For example, it takes between 40,000 and 44,000 pounds of compact discs, about 1.2 million CDs, to make one standard container of shredded, sellable plastic. So while it *feels* as though I have enough promotional AOL discs—they arrive every few weeks in the mail—to make it worth New York City's while to come and collect them, I don't. And neither does anyone else.

When I heard that a Sony Electronics vice president had, at an industry trade show called Waste Expo, proposed dumping electronic waste into open-pit hard-rock mines, I thought he'd been at the minibar too long or was only being poetic: from minerals came these monstrous hybrids and to minerals they would return. But the Sony exec was serious. One mine, he said, would hold seventy-two billion PCs, enough to make the crushing of waste and the extraction of copper, gold, iron, glass, and plastics profitable. Electronic equipment is often richer in rare metals than virgin materials, containing ten to fifty times more copper, as a percentage of weight, than copper ore. A cell phone contains five to ten times more gold than gold ore itself.

While waste traders salivated at this idea, the antimining crowd quivered. Wouldn't deep pits of computers add insult to a system that is already, environmentally speaking, injured? Was this an invitation to dump stuff in a hole and forget about it? Would high-tech miners, wearing biohazard suits instead of Levi's, extract the valuable stuff using cyanide and arsenic, then walk away from what remained? The notion got a little bit of play in the waste world, then cranky recyclers grabbed the microphone and the idea of dumping unholy alliances of metals and plastics into the ground, to be mined another day, slowly sank.

Chapter Nine

Satan's Resin

In the sixties, you could always insult a guy by calling him "plastic." It meant he was phony or superficial. The opposite of plastic was "real." In Mike Nichols's 1967 film *The Graduate*, the hopelessly straight Mr. McGuire, a friend of the Braddock family, offers career advice to the recently graduated Benjamin Braddock. "I just want to say one word to you. Just one word. Are you listening? . . . Plastics." The word became a kind of shorthand for a suburban life of conspicuous consumption and upward striving. It stood for a rejection of old ways and an embrace of modernity, which included the throwaway culture made possible by the expanded use of plastic. Where the *bricoleur* of a century past (that is, an odd-job man who worked with his hands, using the *bricoles,* or odds and ends, that lay at hand), or even Benjamin Braddock's grandparents, had understood metalworking and woodworking, plastic—this wondrous new material—was a mystery. As Susan Strasser writes in *Waste and Want,* "Nobody made plastic at home, hardly anybody understood how it was made, and it usually could not be repaired." Which explained, in part, why there was a pink plastic flashlight pen with a retractable monster tongue sitting in my kitchen waste can.

Across the nation, recovery rates for almost all recyclable materials have declined over the last couple years. But the recovery rate for PET plastic (polyethylene terephthalate, that is, marked by a number 1 surrounded by chasing arrows), the most widely collected type, has fallen especially hard, from a high of 39.7 percent in 1995 to a low of 19.9 percent in 2002, when 3.2 billion pounds of PET bottles were buried or burned. Number one water bottles have an even worse recycling rate than number one soda bottles. In 2002, only 11 percent of plastic water bottles were recycled in the US. And as the market segment grows—and it is growing, faster than any other segment in the US beverage market—the problem is bound to get worse. In 2003, Americans consumed 13 billion liters of bottled water, much of it in half-liter servings, and global bottled-water sales reached 155 billion liters.

Recycling experts link the drop to the rising number of beverages consumed away from home—in offices, parks, cars, and other places that lack a handy recycling bin. The lower recycling rate is a loss for the environment, but it also represents a lost opportunity for PET processors and end users that can't expand their operations or have gone out of business. Had all those bottles been recycled, the Container Recycling Institute reported, "an estimated 6.2 million barrels of crude oil equivalent could have been saved, and over a million tons of greenhouse gas emissions could have been avoided." After e-waste, plastics are the fastest-growing portion of the municipal waste stream: according to the GrassRoots Recycling Network, Americans trash more than forty million plastic soda bottles a day.

Early one morning, I drove out to Farmingdale, Long Island, to see how my yogurt cups, which I had delivered to my local food co-op, were transformed into seawalls and lumber. American Ecoboard sat at the dead end of a bland industrial park, the sort of place where, in heist movies, ne'er-do-wells plan robberies in empty warehouses. I walked through a sad-looking collection of plastic-wood picnic tables and knocked on a metal door. No one answered, so I walked into a vestibule the size of a large port-o-san, calling out a hello. I got no answer, so I let myself in to the inner office.

Calling out "Hello? Hello?" I eventually raised a harried-looking man named Ron Kwiatkowski from a back office. The president and CEO of American Ecoboard, Kwiatkowski was stocky, with a rounded face. He wore blue jeans and a plaid shirt, and he had a mustache and the sort of beard that's mostly shaved, with just a thin line of dark hair around the perimeter of his jaw. We talked for a while about the plastic recycling business—he'd spent ten years working for Coke—and then we walked down a narrow hallway toward the manufacturing floor.

For no good reason I had expected a plastic recycling plant to be filled with bubbling cauldrons of toxic goo. I had imagined white-coated chemists with thermometers in their pockets, test kits at the ready, and beakers lined up on shelves. Instead, Ecoboard's manufacturing floor was dimly lit and populated with large, low-tech machines. It smelled like melting, but not burning, plastic, and the workers, many of whom spoke Spanish, were dressed in jeans and black hoodies. Dust caught in my throat as I watched forklifts scoot gaylords of ground-up plastic across the plant floor. The confetti-sized bits went into a hopper, where they were blended with pigments, anti-inflammatory agents, UV protectors, and fiberglass, for added stability. "It's like making a cake," Kwiatkowski said. "We have a basic model, then we make cakes with different characteristics—reinforced for structural materials, different colors for decking."

After it was mixed for several hours, the batter ran through a series of pipes into extruders, or twenty-five-foot-long tubes, electrically heated to 400 degrees. After the molds were water cooled, an extruder screw pushed the finished product out the end. Beams were cut, just like lumber, and stacked.

"It's very simple," Kwiatkowski said, shrugging. He and his partner had built all this stuff themselves. At first they thought it would take two men to run each of the four lines, but with some tweaking they realized only one worker per line would suffice.

Ecoboard got some of its plastic from groups like food co-ops or local Boy Scouts, but most of it they bought from brokers or MRFs. The company didn't have to buy the odds and ends that showed up on its doorstep, but it paid between fifteen and twenty cents a pound—or four hundred dollars a ton—for the loads they picked up by truck. "This is a pennies business," Kwiatkowski said. "And you can make or lose millions with pennies." The previous year, the pennies had added up to a $3 million profit.

I asked how much plastic Ecoboard used in a year. "In 2003, we'll exceed eight million pounds," Kwiatkowski said. I had pulled a couple yogurt cups, which weighed three-eighths of an ounce each, from my recycling bin for Kwiatkowski, just to leave my mark on this place, but now I felt a little silly adding them to the pile of containers waiting to be ground up.

Before the city cut back on recycling, my plastic (minus the yogurt and cottage cheese cups, which, for complicated reasons having to do with polymer chemistry, were problematic for many recyclers) was picked up once a week by san men from the Brooklyn South 6 and dropped off at a MRF run by Allied Waste in Greenpoint, at the northern reaches of the borough. With only minimal hassle, I got the site manager on the phone. Daren Dutchin immediately set himself apart from all my other sources by inviting me out to tour his facility at my earliest convenience.

I didn't wait more than a day before picking my way through the unfamiliar industrial neighborhood. There were a lot of trucks on the roads here, a lot of honking traffic and diesel exhaust. Scott Avenue was lined with corrugated-metal fences and men in jumpsuits hosing sidewalks. At the avenue's dead end, where the MRF was located, someone had planted a row of linden trees and painted their trunks bright yellow, for safety. One tree lay at a right angle to the sidewalk, severed at truck bumper's height.

Inside Allied's office, the Formica desks were bare and boxes cluttered the floor. Before the recycling suspension, Dutchin had

190 employees handling 550 tons of mixed metal, glass, and plastic a day. Now he had just ten employees, who worked at transferring commercial solid waste from packer trucks to eighteen-wheelers. Once you were established as a waste hauler, it seemed, it was a simple matter to switch your target material. While waiting for Dutchin, a Guyanan with a lilting accent, to get off the phone, I counted no fewer than five wall clocks. Each ticked, but none told the correct time, in any time zone. I read the posters. "Allied Waste Wants You to Improve Our Margin, Protect Our Assets." My favorite safety message said, "Keep in mind: a truck on fire causes low productivity."

Finished with his call, Dutchin led me into the warehouses adjacent to his office. They were dim, oily-floored places with indistinct ceilings. The high windows were broken, and the air was damp. Scores of small blue Dumpsters were clustered together, like a herd of empty ice cube trays. Milk jug caps and flattened juice cartons littered the ground. I shivered and pulled my jacket tighter.

"It was warmer in here when the MRF was running, wasn't it?" I asked Dutchin.

"Not really," he said.

As we strolled through the deserted plant, Dutchin explained the former operation. The packer trucks backed in and tipped their loads of plastic, metal, and glass onto the floor. A grapple pulled out any bulky material and fed the rest onto a conveyor belt that trundled it up, at a forty-five-degree angle, to a trommel. The trommel was a thirty-foot-long rotating horizontal barrel divided into six sections with different-sized holes. Broken glass came out first, dropping onto a conveyor that delivered it to a bunker, or holding bin, then to a hammer mill, where it was pounded into a material that gave recycling proponents agita.

"The glass was nonprofit for us," said Dutchin. "We crushed it and used it at our landfills as alternative daily cover." The cover, which was mandatory, kept down dust and odors and discouraged rats and birds. But because crushed glass was ultimately buried, it allowed the antirecycling crowd to claim that recycling is a waste of time, that all those containers "just end up in the landfill."

Dutchin went on. "The next section of the trommel has bigger holes. Aluminum cans, steel cans, water bottles, that kind of thing, drop to another conveyor." The conveyors here were distant cousins to the rubber-belted things in supermarkets. Five feet wide, they rose ponderously from below the floor, girded with steel bars. The belts rolled through a gallery of eight "pickers," who were paid nine dollars an hour to pluck plastics by type and drop them through crude chutes to bunkers down below. The plastic was crushed and baled, then sold to brokers who resold it, or to mills that shredded the plastic for resale or extruded it, as Kwiatkowski did, into other products. The metal rode for another few feet along the conveyor until magnets pulled out the steel; an eddy current handled the aluminum. Both of these commodities were delivered to Hugo Neu's Greenpoint scrap yard, where they had an appointment with the Prolerizer. The shredded metal was then loaded onto barges, tugged to Jersey City's Claremont Channel, and transferred to ships bound for the highest bidder.

I walked up a metal stairway to the picking line. The conveyor belts were covered with dust and pigeon droppings. In the gloom I made out shards of glass, a floor mat, a dinosaur toy from a fast-food restaurant, and a take-out container stamped with chasing arrows around the number 7. Plastic number 1 (PET, used in soda bottles) and plastic number 2 (high-density polyethylene, or HDPE, used in milk jugs) are the two most commonly collected and recycled types of plastic. Number 3 plastic, polyvinyl chloride (PVC), is used to make pipes, shampoo bottles, carpet backing, and automotive parts. Number 4, low-density polyethylene (LDPE) is used in plastic bags, six-pack rings, and flexible lids. Number 5, polypropylene (PP), appears in bottle caps, snack food wrap, and some containers and film packaging. Number 6, polystyrene (PS), is most commonly found in plastic cutlery and food containers; and number 7, "other," is just what it sounds like, and usually unrecyclable.

The triangular symbol was nothing but a headache for Daren Dutchin. The recycling industry developed it to signify recyclability, but the chasing arrows were appropriated by the Society of the

Plastics Industry in 1988. The arrows make the virgin-plastic manufacturers look good, but they encourage the public to dump anything with a symbol into the recycling bin regardless of whether local MRFs can handle it, along with some plastics, like coolers and sports watches, that have no arrows at all. The result is contaminated loads of material that ultimately have to be dumped. Recyclers have requested that plastic-container manufacturers modify their use of the misleading graphic, but the industry has so far refused.

The vast emptiness of the Scott Avenue MRF gave me the creeps. Not twelve months ago, the warehouse had been a miserable hive of low-wage activity, operating six days a week, twenty-four hours a day. The conditions, I'd heard, were medieval: hot, cold, damp, noisy, dirty, relentless. Now I had the sense that rats were moving stealthily over the rafters, that cats slunk through the derelict equipment, just out of sight. Sumac trees were growing up under the conveyor belts near a train siding. I heard creaking from the roof, the thrumming wing beats of pigeons.

Dutchin, of course, wasn't remotely creeped out by this place. He knew every nut and bolt here, and possibly every rat, too. He was proud of the operation. To him, the emptiness represented lost opportunity: jobs for recent immigrants, a way to reduce the burden on landfills. "It was so beautiful," he said to me now, interrupting my horror movie thoughts. "I wish you could have seen it." Never had two individuals, gazing upon the same scenery, been so out of sync.

I sat down one morning and cut a sheet of paper into small slips. On one I drew a stick figure representing me. On another I drew a horizontal rectangle representing D'Agostino's, my local grocery store. On the third I drew a slightly stylized beverage distribution truck. Then, I cut out two paper nickels and one paper soda bottle, made of PET plastic. And then sliding these scraps around on my desk, I enacted a small play to help me visualize how New York State's bottle bill works. The action began when the distributor brought a cola bottle to the D'Agostino's and the store paid the distributor a nickel deposit. In the second act, I showed up to buy my cola and paid a nickel deposit. I drank the cola, offstage, then returned the empty to D'Agostino's and collected my nickel. In the third act, the distributor returned to the store for the empties and paid back the nickel that set this drama in motion. At this point, the distributor also paid the store a two-cent fee for handling.

But let's change the scenario a bit. What if I bought my cola (or a beer, which also has a five-cent deposit), walked offstage with it, and never came back? Who would keep my nickel? In New York, Connecticut, Oregon, Vermont, and Delaware, it is the distributors. In the six other states with bottle bills, unclaimed deposits go toward recycling education or administration, alcohol treatment programs, or the state's general fund. In 2000, estimates the Container Recycling Institute, beverage distributors in New York retained $140.9 million in unclaimed nickels. Proponents of a bigger, better bottle bill in New York State, which would include sports drinks, water, teas, juices, and other hugely popular "New Age" beverages, are trying to redirect that money—more than $172 million is expected—to recycling and other environmental programs. (Bottle bills in California, Hawaii, and Maine already cover New Age drinks.)

Who could argue with an expanded bottle bill? It would keep litter off the streets and beaches, keep solid waste from the landfill, conserve natural resources through recycling, and direct money to environmental programs. (According to the NRDC, such a bill would lighten New York City's waste stream by 220 tons a day, saving as much as $10 million in curbside collection and disposal costs.) Well, grocery store operators, to name just one group, aren't so keen on an expansion. They'd have to devote more storage space to the sticky, wasp-attracting beverage containers and hire employees to handle them. And then there are distributors, who pay the two cents-per-container fee, and bottlers, who have to clean those containers and find an outlet for them. Waste haulers and MRF owners don't like bottle bills, either: they take weight away from them, and in the garbage world weight equals money. Over the years, packaging, food, and petrochemical industries have quietly spent tens of millions of dollars fighting existing and proposed bottle bills. And they've done it at exactly the same time that they are very publicly promoting recycling.

The garbage landscape is littered with greenwash tactics, in which polluters pose as friends of the environment but spend more money advertising their green projects than on the projects themselves. One masterful example of corporate greenwash is the Keep America Beautiful campaign, which was founded by beverage companies and packaging executives in 1953 after magazine ads began promoting beverage cans as "throwaways" (one depicted carefree boaters slinging empties into a lake). Litter alongside roads, rivers, and farm fields had begun to accumulate, prompting Vermont to pass the nation's first bottle bill, which banned the sale of beer in nonrefillable containers. Beer companies didn't like that one bit. They lobbied hard against the law, and in four years it expired. (The state enacted a new bottle bill in 1972.)

In a stroke of marketing genius, Keep America Beautiful (KAB) urged individuals to take responsibility for this waste, to "put litter in its place." In 1971, the organization sponsored one of the most successful public-service announcements in history, a TV commercial in which a Native American, complete with braid and eagle feather, paddles down a pristine waterway until he reaches a teeming city. When he spots empty beverage cans swirling in the shallows, a tear rolls down his leathery face. KAB proudly called the "Crying Indian" spot an "iconic symbol of environmental responsibility." (Iron Eyes Cody, who played the Indian and claimed to be a Cherokee-Cree, was later outed as a Sicilian American named Espera DeCorti.)

But whose responsibility is the foul mess along the shore? The organization's underlying message is that individuals, not corporations who produce single-use containers, are responsible for trash, and that individuals must change their behavior, not manufacturers. Keep American Beautiful focuses on antilitter campaigns—which enlist millions of volunteers a year to clean up beaches and roadsides—but it ignores the potential of recycling legislation and resists changes to packaging. Between November 1992 and July

1993, the American Plastics Council, a KAB sponsor, spent $18 million on a national campaign to "Take Another Look at Plastics." The ads crowed that more than a billion pounds of plastic had been recycled in 1993, but they failed to mention that fifteen billion pounds of virgin plastic were produced during that same eight-month period. According to a report by the Environmental Defense Fund, for every one-ton increase in plastic recycling between 1995 and 1996, there was a fourteen-ton increase in new plastic production.

For twenty years environmental groups, including the Sierra Club, the National Audubon Society, and the National Wildlife Federation, lent legitimacy to KAB by sitting on its advisory committee. Those relationships ended after a board meeting in July of 1976, when American Can Company chairman William F. May denounced bottle bill proponents as communists and called for a total KAB mobilization against proposed bottle bills in four states. Today, KAB is funded by about two hundred companies that manufacture and distribute aluminum cans, paper products, and plastic and glass containers, in addition to companies that landfill and incinerate all of the above.

Do bottle bills work? Do they "put litter in its place"? According to the Container Recycling Institute, the eleven states with bottle bills recycle beverage containers at a rate of 70 to 95 percent, while states without bottle bills average 37 percent. (Though, thanks to the declining value of the nickel, and, again, Americans' increasingly mobile lifestyle, the percentage of cans and bottles redeemed even in bottle bill states is dropping.) New York's bottle bill is estimated to divert more than 650,000 tons of aluminum cans, and glass and plastic bottles from the state's municipal waste stream each year. Of states that track how much waste their container deposit laws divert, Iowa clocks in with 50,000 tons, Maine 54,000 tons, and Vermont nearly 16,000 tons.

There is a lot of green pride in New York over the bottle bill, but it isn't widely known that the law doesn't require distributors and bottlers to actually recycle the containers they collect back into new beverage containers. Only between 4 and 6 percent of glass bottles sold in New York are refillable; the rest of the redeemed bottles are melted down and reincarnated as new containers. It is technologically possible to blend recycled plastic with new resin in plastic bottles, but there isn't much incentive to do so. (In Scandinavia, Germany, and the Netherlands, consumers return PET bottles, which have a sturdier formulation in those countries, to companies that sterilize and refill them again and again.) In the United States, packagers and manufacturers prefer new plastic because it is cheaper than recycled plastic, it's free of incompatible polymers, and its color is easier to control. The quality of virgin plastic is guaranteed, and the infrastructure to make it is already in place. Pressured by shareholders in 1990, both Pepsi and Coca-Cola promised to use 25 percent recycled plastic in their bottles, but neither company did. In 2000, both companies committed to using 10 percent recycled content by 2005.

So where are all the postconsumer plastics going? If they are part of a mixed stream collected at curbside, they might end up at a place like American Ecoboard. If they are PET bottles redeemed in bottle bill states—and therefore cleaner and less contaminated with other grades of plastic than loads of curbside bottles—they are probably transformed into sleeping-bag fiberfill, carpets, and fleece jackets. Turning Sprite bottles into Synchilla hoodies is slightly more complicated than turning shredded pellets into picnic tables. The bottles are first shipped to a processing plant to be washed and granulated. The flakes are dried and sold to a mill, where they are melted and squeezed through tiny holes in flat plates called spinnerets (named for the tubular structures from which spiders secrete silk threads). After the plastic solidifies, it's spun into long threads and stretched to many times the fiber's original length. The strands are then crimped into wave patterns, using heat, and cut to length, ready for weaving into Synchilla, Capilene, Polartec, or another product with a fuzzy-sounding name.

Not all postconsumer plastic cycles back into gross domestic product. According to the Association of Postconsumer Plastic Recyclers, 35 percent of PET bottles collected in the United States in

2003 were exported, mostly to China. The bottles followed a national trend: more and more, recyclable materials for which industry has no use or that it can't afford to process are being sent overseas. In 2002, we shipped to China about 450,000 metric tons of scrapped plastic (more than seven times the amount in 1996), 3.3 million metric tons of paper (more than five times the 1996 figure), and 2.3 million metric tons of scrap iron and steel (nine times more than in 1996). This global trade sends recycling jobs overseas, but it gives us cheap goods. Is it a fair deal? Our nation consumes more than its share of natural resources, we create the most waste, and then we send it to be processed in countries that fail to protect their workers or their environment from industrial pollution. Sure, overseas workers get jobs, but they also get contaminated water, soil, and air. Seattle's recycling program owes much of its success—it had a 39.7 percent diversion rate in 2002—to strong overseas markets for plastic. But that could change. "People are starting to realize that recycling isn't so simple," said Pete Erickson, of Seattle's Cascadia Consulting. "We're thinking about our impact overseas. We want to be a good global citizen."

In the midnineties, Greenpeace researchers poring over US Customs Department data discovered that the Pepsi-Cola Company was shipping plastic scrap to Madras, India. When the company denied the practice, Ann Leonard, who now works at the Global Anti-Incinerator Alliance, packed her bags and went off in search of proof.

"When I first got to Madras, I went around to all the ragpickers, because they know everything," said Leonard, who speaks in rapid-fire bursts. "They're very organized and hierarchical. Each family does a different resin: one does only PET; another does only HDPE. The ragpickers were mad because they couldn't compete with all the plastics coming in from overseas. They were losing their livelihood." The ragpickers gleaned plastic from roadsides and trash cans, selling it to small factories that made low-quality plastics for sandals and kitchen goods. "You know," said Leonard, "all that stuff you see in Third World countries that breaks. Cups and bowls."

"And toys?" I asked.

"Yes!" Leonard yelped. We both had young children, and every time they went to birthday parties—her son on the West Coast and my daughter on the East—they came home flush with candy and cheap plastic toys made in Asia. "And they always end up in the trash," Leonard said, sighing.

I'd never talked to anyone so vehement about plastics. A native of the Pacific Northwest, Leonard had always thought she'd work as a forest activist. Then she went to school in New York and was bowled over by the amount of cardboard piled on sidewalks for collection. "That's where all our forests are going," she said to herself. Almost overnight, she dedicated herself to waste activism instead.

Riding around Madras in a rickshaw, Leonard ordered her driver to a halt every time she spotted a plastic bottle in a ditch. Over and over again, she saw bottles stamped "California Redemption Value"—bottle-bill bottles! "Finally, we came over a rise in the road and saw this enormous pile of compressed and baled bottles. The factory, owned by Futura Industries, washed, chipped, and melted the plastic, which it added to virgin plastic to spin polyester fabric. Out back there was a pile of waste—the hard bottoms of PET bottles, their lids and labels." According to the plant manager, Futura processed only between 60 and 70 percent of the bottles it received (not unlike Allied, back in Brooklyn). The rest were either too contaminated with residual material, with other garbage that arrived mixed in with the shipment, or with substances impossible to recycle. This last category was growing by leaps and bounds as bottlers introduced weird hybrids into the marketplace: plastic bottles with aluminum tops, tinted or painted plastic, and bottles made of multiple layers.

According to Greenpeace, 50 percent of the discards shipped overseas were contaminated. Importers were left with mounds of plastic that they either dumped on the ground—often in unlined, unmanaged sites where they leached toxins into the soil and water—or burned. Greenpeace reported that none of the recycling workers employed by Futura—30 percent were women earning less than thirty cents a day, 60 percent were children, and the remaining 10 percent were old and disabled men—wore a mask or other clothing that would protect them from noxious fumes released by burning plastic, a combination of carbon dioxide, carbon monoxide, and sulfur dioxide, and, in the case of PVC, dioxin. Dioxin migrates on the wind, settling on grasslands and in the water, where grazing animals and fish consume it. Like DDT, dioxin doesn't readily break down in fatty tissues: it accumulates. According to medical researchers, traces of dioxin can be found in every person on earth.

In response to Greenpeace's report on the dumping situation, Pepsi claimed that workers were not endangered, that Coke was doing the same thing, and that this was what bottle recycling looked like. "But it's not recycling at all," said Leonard. "True closed-loop recycling has no new resource input and no waste output. And that's virtually impossible with plastic waste because its chemical structure changes when it's heated and the quality degrades. We're just delaying its eventual dumping."

I hung up the phone and stared out the window for a minute. If Leonard was right, then it didn't matter whether I redeemed my plastic bottles at the store, the first step on a journey to Asia, or gave them to American Ecoboard, via my food co-op. "Recycling" plastic, because it created new toxins and left old ones behind, might be more harmful than landfilling.

Leonard had suggested I call Berkeley's Ecology Center, which developed the East Bay's highly successful curbside recycling program, fueled its collection trucks with biodiesel, and ran a storefront that sold environmentally friendly carpet shampoo, compost bioactivator, whale magnets, and relaxation tapes. But for all its orthodoxy, the center for many years refused to collect plastic, which its founder, operating on the same wavelength as Ann Leonard, preferred to call "Satan's resin." Why such opposition? Because picking up plastic at the curb, said the Ecology Center, would legitimize the production and marketing of packaging made from virgin plastic, imply that it was ecologically friendly, and encourage residents to buy more of it. Alas, all this abandoned Berkeley plastic would only end up in the landfill. (In 2001, the city began to collect number one and number two containers, but only if they had necks narrower than their bases.)

As Leonard said, plastic isn't truly recyclable in the way that glass, metals, and fibers are. Streams of mixed plastic can be turned into only one other product (plastic wood, garden pavers, or toothbrush handles, for example). When their useful life is over, these products cannot be "recycled" again. They have to be burned or buried. Either way, they add toxins to the environment. Unmixed streams are another matter: they actually can be refashioned into bottles and containers. But there isn't much demand from their makers for recycled plastic. Virgin is so much cheaper.

And even if plastic manufacturers magically got it together and began using recycled content, the Ecology Center would still take issue. The raw material for the plastic used in packaging is ethylene, a gas derived from natural gas or from a fraction of crude oil that has a composition similar to natural gas. "Both natural gas and crude oil are products of fossils and are therefore not renewable," says the Berkeley Plastics Task Force report.

Producing and refining ethylene is a multistep process, one that employs small armies of those white-coated chemists I mistakenly conjured at American Ecoboard. First, the gas has to be heated, then refrigerated, then combined with solvents, comonomers, additives, and other chemicals. The mixture is then "polymerized" to create long-chain molecules. The new polymer is extruded, pelletized, or flaked: the finished product is called a resin. The resin is sold, reextruded, and made into containers, films, and other products.

If it sounds energy intensive, it is. But even worse, plastic is toxic both to make and to dispose of. On the front end, says the EPA, the production of plastic emits the toxins trichloroethane, acetone, methylene chloride, methyl ethyl ketone, styrene, toluene, and 1, 1, 1 trichloroethane, as well as sulfur oxides, nitrous oxides, methanol, ethylene oxide, and volatile organic compounds. Plastic manufacturers use copious quantities of benzene and vinyl chloride, which are known to cause cancer in humans. Ingesting other ingredients of plastic production can lead to birth defects and damage the nervous system, blood, kidneys, and immune system. Many of these chemicals are gases and liquid hydrocarbons that readily vaporize and pollute the air; many are flammable and explosive, and many can cause serious damage to ecosystems. In an EPA ranking of the twenty chemicals whose production generates the most total hazardous waste, five of the top six are chemicals commonly used by the plastics industry. Not surprisingly, plastic resin factories tend to be clustered in low-income communities of color (mostly in the Gulf States, which have easier access to gas lines). OSHA health studies have shown that people who work in and live near plants that manufacture plastics and the chemicals used to make them experience higher incidences of some kinds of cancer than other populations.

At the end of their useful lives, plastic products that lie by the roadside or get buried in landfills can leach phthalates—which give plastic its softness and flexibility but have been linked with endocrine disruption—into groundwater. Burned in an incinerator, shampoo bottles, take-out containers, and bathtub mats release other toxins that escape smokestacks or are concentrated in bottom ash, which is eventually buried in landfills (unless it is combined with other materials and used in construction).

Of all the materials we throw out, plastic is among the hardest to kill. It doesn't biodegrade in any conventional sense; sunlight causes it to photodegrade into ever-smaller pieces of polymers. These are easily consumed by some organisms, but they're still too large and too tough to be digested by microorganisms. In a landfill, where the sun never shines, plastic doesn't get even this far. ("Earth friendly" biodegradable plastics, made of potato- and cornstarch, need moisture to break down; this, too, is in short supply within most landfills.) But washed into the ocean from rivers and streams, dropped overboard from boats, or abandoned as fishing nets, plastic degrades into pieces that choke turtles, entangle jellyfish, and fill the stomachs of seabirds from the tropics to the antipodes, which then starve to death because they always feel full.

Besides the usual bits of balloons and bags, Laysan albatross chicks have ingested a cigarette lighter, a toothbrush, a tampon applicator, a toy robot, a golf ball, and lids from a car battery and a shampoo bottle. In 1999, marine researcher Charles Moore surveyed five hundred square miles of the North Pacific subtropical gyre and found six pounds of floating plastic for every pound of naturally occurring zooplankton. He repeated his study in 2002 and found ten pounds of plastic for each pound of zooplankton. A 2004 study conducted by marine ecologists around the British Isles showed accumulations of microscopic fibers and bits of synthetic polymers in beach and seabed sediments, as well as a big jump, in the last two decades, in the concentration of plastic particles amid plankton.

The more I learned about plastic, the worse I felt about the way I transported short-grain brown rice from the food co-op to my home (in a number 4 LDPE bag that I reused) and stored my leftovers in the fridge (in number 5 polypropylene containers). Not only was plastic bad news, both coming and going, but trying to recycle it possibly made the situation even worse. "It's just a diversion from more important issues, like sending putrescibles—very valuable stuff—to the landfill," Dan Knapp told me. Knapp was part of the Berkeley Plastics Task Force, and he ran that city's Urban Ore, a reuse and recycling center that kept five thousand tons of "waste," in hundreds of different categories, in circulation and out of the landfill. "We should just ban plastics. They're not worth it."

After talking to Knapp I reviewed my own garbage data. It's estimated that Americans go through about a hundred billion polyethylene bags—the ubiquitous eighteen-microns-thick grocery sacks that snag on branches, skip along on the breeze, clog sewers and storm drains, and burrow into ditches and dunes—a year. Although plastic bags don't take up a lot of landfill space, they persist in the environment for decades, if not centuries. Like other forms of plastic, they have high social and environmental costs—called "externalities"—that are borne by the public and by government, not by the producers of the plastics or their intended users. Recognizing these externalities, South Africa has prohibited the sale of plastic bags under 80 microns thick, and Taiwan and Bangladesh, where plastic trash clogged street drains that carried human waste, have banned free distribution of the bags in stores. Ireland reduced bag use by 90 percent by instituting a fifteen-cent charge for each sack.

Because they were so light, plastics left barely a mark in my trash logs, though I was going through an average of 5.2 Ziplocs and thin vegetable bags a week. When I began separating the bags from my kitchen trash, the total number of items in the can fell by nearly half. I ignored the slimiest bags, but the torn veggie bags, the worn-out Ziplocs, excess shopping sacks, pretzel and spinach and cheese bags, scraps of Saran Wrap, bread bags, and their ridiculous inner plastic liners now collected in yet another bin in my personal materials recovery facility (my kitchen). After one month, I had an entire pound of them.

Until producers took back the resins they sent out—I figured this would take a legislative act—I was going to have to change my habits. Instead of carrying my brown rice home in a plastic bag, I could buy it in a recyclable box. That sounded good until I considered its product-to-package ratio. According to California's Integrated Waste Management Board, a delivery of one thousand pounds of rice in plastic bags generated 3.9 pounds of waste, while the same amount of rice delivered in paperboard generated 78.1 pounds of waste. Which was preferable? The choices, like so many at the intersection of consumerism and environmental concern, were agonizing.

Switching from bottles of liquid dish soap to cakes of hard yellow soap, which worked great and came with zero packaging, was a no-brainer. I was already reusing my Ziplocs, but I resolved to always use containers, rather than Saran Wrap, to hold leftovers. I checked my data sheets again: the only other plastics that occurred in my trash were bottles of shampoo, conditioner, olive oil, ketchup, mouthwash, medicine, and, twice, children's bubbles. A year's data included three half-liter water bottles, but on that matter my conscience was clear: they were outliers, introduced by guests unaware of my single-use phobia. I was devoted to my wide-mouthed Nalgene bottle—refillable, hardy at all temperatures, a cinch to clean. Then I read about a study conducted at Case Western Reserve University and learned it was made of a polycarbonate called Lexan that's been linked in mice to an endocrine disruptor called bisphenol-A, which has in turn been linked to chromosome abnormalities and the runaway development of fat cells. The only healthy alternatives for toting around liquids, it seemed, were the leather bota bag, popularized by Chianti-drinking campers in the seventies, and the bladders of large ungulates, like buffalo or elk, popularized by hard-core survivalists.

But what about the plastic bottles I used at home? I decided to buy ketchup only in glass. I would buy olive oil in cans, then give them to Wendy Neu. I could buy shampoo and conditioner in the largest-size plastic bottle I could find. It was either that or go for those antiquey-looking products, usually "botanical," that came in blue glass. But they were expensive, and heavy, and slippery when wet, and impossible to squeeze the last and even second-to-last drops from.

Exactly, I could hear the plastics industry murmuring as I made the case for Satan's resin.

Part Three

Flushing It Away

Chapter Ten

Downstream

My garbage had a peculiar smell, a smell it hadn't had in two years. It wasn't the slimy vegetable bags or the moldy sour cream, which I should have flushed down the toilet, but a tightly wrapped 9.5-ounce plastic triangle of human waste, courtesy of a visiting three-year-old. I'd been feeling fairly good about my garbage recently, but all of a sudden it had plunged back into the offensive zone. It occurred to me that I never would have started quantifying my garbage if my daughter were still in diapers. Her embrace of underpants was not only a benchmark along the path of child development, it was also a banner day for my garbage. I wasn't sure which aspect of the event thrilled me more.

Disposable diapers occupy a special place in the cultural history of landfills. As the burning Cuyahoga River woke the nation to the deplorable state of our urban waterways, the perception that disposable diapers were clogging our dumps acted as a clarion call in the late eighties. Whether disposable diapers actually *were* busting out of landfills didn't matter: they made a convenient symbol. Because they saved parents time and gave them freedom, they underscored our laziness and selfishness. Because they were filled with shit, they added currency to our general disgust with garbage. The alarm sparked a wave of cradle-to-grave studies evaluating the environmental impact of disposables versus reuseables. The conclusions fell, predictably, along the lines of who funded the research.

The diaper issue continues to absorb a good bit of researchers' energy, but EPA data reveal what may be a tempest in a teapot. Disposable diapers (for both infants and adults) constituted 1.9 percent, by weight, of total US landfill discards in 1995. In 1996 and 1997, the percentage held steady at 2.0 percent, then rose in 1998, the last year for which data are available, to 2.1, when 3.4 million tons of disposables were buried. Were Pampers, which according to the pro-cloth lobby could take up to five hundred years to decompose, poisoning an otherwise pristine environment? Hardly. The Garbage Project noted that landfills already receive about 20 percent of the sludge from America's sewage treatment plants. This last factoid gave me pause: maybe it didn't matter whether I put my used tissue in the trash or in the toilet: it all ended up in the same place. Or did it?

Since I was traveling with different kinds of things that I disposed of, it was only fair, I reasoned, to follow that tissue down the bathroom pipes. This particular journey began when Phil Heckler, who'd recently retired from the city's Department of Environmental Protection, double-parked his blue minivan outside my house early one Sunday morning and pushed open his passenger door.

The first thing Heckler did, after we shook hands, was open a thick roll of inflow and infiltration maps. He pointed to the spot where my private effluent moved into the public domain—more or less right under the van. The maps showed the street grids and block numbers, the dimension of pipes, the location of tide gates and diversion and regulator chambers, the depths of pipes, and the elevation, below mean high water, of sewage interceptors. "Let's see where your line starts," he said, shifting the van into drive.

Heckler pulled over where Seventh Street ended, at Prospect Park West. The Brooklyn underworld was guarded at this spot,

Cerberus style, by a triumvirate of utility manholes: one belonged to the DEP, representing sewers and water; one belonged to ConEd, in charge of electricity; and one belonged to Keyspan, the local god of natural gas. Under the DEP manhole lay a sewage pipe cast sometime in the late 1800s of vitrified clay. At the start of the line, the pipe was just twelve inches in diameter, an insignificant twig compared to the mighty branch it would eventually become. Within half a block of my house, the pipe would grow by three inches; before it ran into the treatment plant, three miles away, the pipe would be nine feet in diameter.

"Sewer pipes are designed to flow at a rate of two to three feet per second," Heckler noted in his quiet, patient way. He tended to explain mechanisms with drawings, and I liked that. He had a scientist's precision but was careful not to overexplain. "You need a bigger sewer when the terrain is flat. You don't want the material to settle out." "Material" was anything that went out of the house in a pipe, flowed into the system from a storm drain, or trickled in through cracks. Heckler began writing, in engineer-neat script, *V* for velocity, *Q* for flow settling, and *S* for steepness—variables that determined appropriate pipe diameter.

A civil engineer, Heckler had spent twenty-seven years working for the DEP, an enormous agency with tentacles that reached beyond the city limits to upstate watersheds and downstate beaches. The DEP's Bureau of Wastewater Treatment had 2,000 employees who operated 14 sewage treatment plants, 89 pumping stations, 8 dewatering facilities, 490 sewer regulators, and 6,000 miles of intercepting sewer pipes. They also tested water quality at 80-odd monitoring stations along 425 miles of shoreline and inspected and cleaned more than 137,000 catch basins (it took three years to make the rounds of them all).

Before we met in person, Heckler and I had talked on the phone about my catchment. We assumed that my sewage flowed west and was then pumped north to the Red Hook treatment plant, which wasn't actually in Red Hook, but in Williamsburg, a little farther north. Before that plant went on line in the 1980s, Park Slope's sewage had run directly into the Gowanus Canal. In fact, our maps still indicated that there were plenty of pipes discharging into the canal. "It was very common to end the line at the water instead of sending it into the system," Heckler said.

I'd been visiting the Gowanus off and on for some time now, on my own and with community groups. At low water, we dragged canoes over rotted wooden bulkheads impregnated with raw sewage. At high water, it was easy to imagine that oysters could once again flourish here, even if you'd never in a million years want to eat them. I felt geographically connected to the Gowanus because I lived on a hill and the canal was the low point on my western flank. My raw sewage, if I'd lived here twenty years ago, would have settled into the muck at the canal's bottom. I was explaining all this to Heckler when he interrupted my reverie of downstream interconnectedness.

"Your effluent doesn't flow into the Red Hook plant," he said. Instead, it took a sudden turn on Third Avenue and ran south to Owls Head, a facility at the southern tip of Brooklyn. And so to Owls Head we went, following the unidirectional arrows on the inflow and infiltration map to Sixth Avenue, where the pipe hopped up to eighteen inches, then on to Fifth and then Fourth Avenue, where the pipe leaped to seventy-eight inches as the sewage of my many neighbors commingled with mine.

At Third Avenue we turned south, driving in the shadow of the Brooklyn-Queens Expressway. The neighborhood was more commercial here, filled with businesses dumping bad stuff into their pipes and sewers. From gas stations came brake and transmission fluids, from restaurants came grease. These establishments were required to install grease traps that they regularly emptied, but violators were plentiful and came from every social stratum. Domino's Pizza made the DEP's list of Significant Noncompliance; so did the swanky Tavern on the Green. A little more than a century ago, grease and fat were hot commodities that factories turned into candles, soap, and lubricants. During World War II, Americans were exhorted to save kitchen grease for explosives. I called American Waste Products, a local service provider, to see what became of the french fry oil they collected from Brooklyn restaurants.

Like so many others in waste management, the company exercised its right to remain silent. I had no luck with two other grease recyclers, then reached an employee at Filta-Clean who admitted they offloaded their stuff at A&L Cesspool Service, in Queens. "I think they turn it into soap and cologne," he said in a tone that reeked of Male Answer Syndrome. When I asked A&L to confirm this, they hung up on me.

Heckler thought recycled grease had something to do with lipstick. I had hoped for something more *en vogue*: biodiesel. Alternative-energy experts estimate that forty million gallons of this nonpolluting fuel, which can be made from virgin soybean, corn, canola, coconut, or peanut oil, or by filtering and processing used vegetable oils, courses through the combustion systems of retrofitted vehicles nationwide. (And, of course, through the garbage and recycling trucks of Berkeley, California.) Biodiesel is substantially cleaner than regular diesel. According to a 1998 study by the National Renewable Energy Laboratory, it reduces emissions of carbon monoxide by 43 percent, hydrocarbons by 56 percent, particulates by 55 percent, and sulfurs by 100 percent.

My brother, who lives in Maine, introduced me to a friend who drives a 2001 Golf that smells, just faintly, of french fries. "I collect grease at four local restaurants," Don Hudson, its owner, told me. "We make fuel in the warm part of the year, when the restaurants are busy with tourists, then we use it year 'round in the Golf, in an old Volvo, and in two fifteen-passenger vans." The vans and the Volvo, which sport bumper stickers of a large soybean dripping oil, belong to Hudson's Chewonki Foundation, which focuses on environmental sustainability and prefers its biodiesel "neat"—that is, unsullied by any fraction of fossil diesel. Producing the stuff is simple: in a barn, Chewonkians heat the grease on a glycerin-burning stove, then mix it with sodium hydroxide, a.k.a. lye, and ethanol, which they make from local corn. "Converting to biodiesel was the quickest way to lower the foundation's total carbon emissions," Hudson said. "We'd already done all the insulating of buildings we could do. Tackling transportation was obvious." By Hudson's reckoning, he'd kept fifteen thousand gallons of fry oil out of landfills so far. As his operation expanded, with a partner, he aimed to convert one million gallons a year. It was a heartening thought: so long as human beings ate fried food, biodiesel would qualify as an alternative energy source that was sustainable.

Heckler and I passed a squat brick building surrounded by an iron fence. The place looked derelict, with tall weeds and graffiti, but Heckler assured me its pumps were busy hoisting sewage up from fourteen feet below street level to minus two feet, then sending it on its way again. At Twenty-eighth Street, the pipes graduated from 78 to 108 inches in diameter. "We picked up another sewer," Heckler said, examining the map. The new line came from the Metropolitan Detention Center, a bleak-looking federal prison that rises between the Brooklyn-Queens Expressway and New York Harbor. "There have been some problems here," Heckler said. His concern was "extraneous material" being flushed down the toilets. "The inmates flush rags and ketchup packets," he said. "It's their way of protesting." We drove another mile and turned onto First Avenue, where, eight feet down, my 108-inch pipe and another 60-inch pipe fed into a regulator. During a hard rain, this contraption directed my untreated sewage straight out into the harbor.

Humans have done all kinds of interesting things with their waste through the ages, including burning it for fuel and composting it, but sewage treatment itself is a fairly young science. It wasn't until the midnineteenth century that European health officials made the mental connection between sanitation and public health. Engineers in modern cities had already laid pipes to shunt storm water to rivers; it was a small leap to envision the same pipes carrying waterborne sewage. Until the 1930s, all New York pipes ended at waterways; so did pipes in Boston, Philadelphia, Atlanta, and around the Great Lakes. As populations grew and the fouled waters became hard to ignore, treatment plants began to come on line. But it was a slow process. As recently as the seventies, New York was still discharging 450 million gallons of raw sewage a day into the waterways surrounding the five boroughs. Until 1986, the entire west side of Manhattan, north of Canal Street, discharged its sewage into the Hudson.

Improvements to the system decreased discharges, but they have never ceased. During heavy rainstorms, runoff from the street joins sewage in the pipes and overwhelms the system: the excess is shunted, untreated, through a local regulator and out into the rivers. In a big storm, nearly 40 percent of the flow that enters the city's sewer system exits that system untreated. "The system is called a CSO, for combined sewer overflow," Heckler said. "And most large, old cities have them."

According to a report from the NRDC and the Environmental Integrity Project, older sewer systems in the Northeast and around the Great Lakes dump an estimated 1.3 trillion gallons of raw sewage into community waterways each year. In Hamilton County, Ohio, a single sewer annually discharges as much as seventy-five million gallons of untreated sewage into Mill Creek, including during summer months, when children swim in the river. As little as a half-inch of rain in Washington, D.C., can cause sewers to overflow into the Anacostia River, which runs through the heart of the city. In Indianapolis and surrounding Marion County, CSOs occur sixty-five days a year, discharging a total of seven billion gallons of raw sewage into the White River. In 2001, Michigan reported 463 CSO events, dumping a total of thirty-one billion gallons of sewage into state waterways.

About 490 regulators dot New York City's waterfront. That is 490 places where combined sewage outflow joins the waterways. An additional 250 outfalls dump only storm runoff, which doesn't sound too bad until you consider that rainwater scours from city streets and parking lots a toxic stew of polycyclic aromatic hydrocarbons—a group of more than a hundred chemicals formed during incomplete combustion—industrial metals, volatile organic chemicals, motor oil, copper from brake linings, lead from paint, zinc from the corrosion of galvanized steel, illegally dumped restaurant grease, and a liberal peppering of street litter and dog droppings. Heckler tried to be reassuring: "We made huge strides by making small changes in the system—removing bottlenecks and putting more people on in wet weather to pick out debris and monitor the flow. In 1930, we captured zero percent of the city's wet-weather flow. In 1987, we captured eighteen percent. Now we're at sixty-two." You had to take his word for it.

We continued south toward Bay Ridge, scanning the horizon for gulls. Heckler wasn't sure where the Owls Head treatment plant was—his map didn't cover it—but the birds find plenty to scavenge around primary treatment tanks and so make excellent mobile beacons.

The gulls led us to a collection of ochre-colored concrete buildings that jutted from the shoulder of the south Brooklyn shoreline. Owls Head was oriented toward the open harbor and afforded great views, from the poop deck of its settling tanks, of the Verrazano and Brooklyn Bridges, the Statute of Liberty, and the southern tip of Manhattan. In aerial photographs, Owls Head looks slightly military, like an aircraft carrier but paved with treatment tanks instead of runways. A splotchy cloud of brown scum floated in the water near the plant, which was just a couple hundred yards north of a fishing pier. I once asked some anglers there if they knew what went on in the plant and whether they ate the bluefish and stripers they caught. The answers were yes and yes. "If the fish's tongue is black, I throw it back," one of them told me.

Owls Head had been built in the 1940s and in 1995 upgraded to meet the Clean Water Act's "secondary treatment" standards, which meant the plant removed 85 percent of biochemical oxygen demand—a measure of how much oxygen is being consumed by microorganisms—and discharged fewer than thirty milligrams of suspended solids per liter of water. Every day, 120 million gallons of sewage flowed into the plant and was separated, through a process of settling and digestion, into two physical states: liquid and solid.

Heckler flashed his ID at the plant's entryway and parked the minivan around the back of a low building, in the shadow of several concrete tanks. We let ourselves into a long, narrow room filled with control panels, computer screens, switches, dials, and a lot of green and red lights. "How you doing?" Heckler said, reaching for the outstretched hand of Lou Gibaldi, a jowly engineer dressed for Sunday in a sweatshirt and two days' stubble. Heckler introduced me, then asked, "Before we get started, could I use the facilities?"

Left alone for a moment in the electric control room, I pondered the question: what would "the facilities" look like in a place like this? Like a litter basket in the cab of a garbage truck, a toilet in a wastewater treatment plant seemed superfluous. Heckler reappeared, and we went off to explore. We strolled past a pump and an eight-cylinder engine, which ran on homegrown methane, then came upon a forty-foot-deep chamber set into the concrete floor. Grayish green water coursed through at a rate of two feet per second. It was an impressive sight—this raw effluent so casually racing by, the gash in the floor surrounded by the sparest of iron railings. I thought of the Augean stable, filled with thousands of cows. On orders to muck the place out after thirty years' neglect, Hercules cleverly bent two rivers to flow through the stable and swept the filth straight out. New York's forefathers envisioned a similar system a century and a half ago when they diverted pure drinking water from upstate rivers, via aqueducts and reservoirs, and sent it through the city's pipes and then out again—much, much dirtier—into the harbor.

I watched as a coarse screen and then a fine screen lifted debris from the sluiceway and deposited it in a Dumpster, bound for the landfill. I spotted apple cores, tampon applicators, condoms, and what appeared to be a calamata olive stuck to the grate. Though relatively few New Yorkers have disposals, a fair amount of food ends up in the system anyway. Just last week, I'd dumped some spoiled yogurt into the toilet: dairy couldn't go into the compost, the cup was recyclable at the food co-op, and I didn't want to slime the garbage in my kitchen trash can, knowing that I'd soon be picking through it.

"I don't think that's an olive," Heckler said.

"I guess not," I admitted, as the reality of exactly where we stood sunk in.

From the grate, the sewage flowed into a thirty-foot well; five pumps lifted the flow up and out into four primary settling tanks outdoors, where it stayed for an average of one hour and six minutes, or 1.1 hours in engineer speak. "You get settling action in the primary tanks," Heckler explained as we walked across the parking lot and up a short flight of stairs to the long rectangular pools. The water that circulated here was flat and brown, except where plastic straws, coffee stirrers, pens, caps, and more tampon applicators clustered in an off-white froth. Gulls pecked at the edges of the tanks; the air smelled of hydrogen sulfide. It wasn't too bad on a breezy winter afternoon; I imagined things were probably a lot worse in the dog days of August. Wooden flights—long thin planks of redwood—slowly skimmed across the top of the pools, pushing grease to one end, then dove down fifteen feet and traveled in the other direction, scooping solids to the next stage of processing.

"Where does the stuff from on top go?" I asked.

"To the scum concentrator," Heckler answered. He described a machine that squeezed out water and reduced grease to the density of congealed bacon fat. The concentrated scum, larded with bits of plastic, was sent to the landfill. Just the name of the contraption nudged me over my personal gross-out threshold, and I was almost grateful that Heckler didn't have time to show it to me.

One of the settling tanks was empty, with a long-handled scrub brush leaning against its blue-painted side. Hand-cleaning these encrusted tanks was probably one of the worst jobs in New York. When Gibaldi first came to Owls Head, he had told me, he was on the bottom-most rung of the status ladder: he cleaned. Seventeen years passed, and now Gibaldi was philosophical about his career. He enjoyed giving tours to schoolchildren. "I tell them whatever they see here started out as clean water." He smiled earnestly. "You know, if this place wasn't here you could *walk* to Staten Island." It took me a second to grok what he meant, that New York Harbor, in the absence of sewage treatment plants, would eventually be solid with human waste. Gibaldi's transparency was a breath of fresh air, and it made me wish that landfill managers were equally enlightened. I had a feeling, though, that the issues surrounding sewage treatment were far less complicated, at least politically, than those that swirled around trash. Soon this feeling would fade.

"I tell the kids to think of the sewage treatment plant as a giant human body," Gibaldi continued. "It's a digester, really." He explained how the same microbes that attack food in our stomach and intestines, that break it down into simple substances and create methane, attacked sewage at the plant. Owls Head was simply concentrating and accelerating the decomposition that would naturally take place.

Heckler and I moved from the primary tanks to the aeration tanks, where the bacteria that had ridden in with the waste, enlivened by a jet of air, ran wild with the food and feces. When sated, these microbes fell to the bottom. As the solid waste settled out of the liquid, it was drawn to final settling tanks, where heavier particles and other solids again sank to the bottom and more liquid was drawn off. A small part of this sludge was recirculated to the aeration tanks as "seed" to stimulate digestion of the newly arrived waste; the remaining solids moved inside the plant, to thickening tanks, for further processing. The wastewater extracted at each stage was treated with sodium hypochlorite—a.k.a. liquid bleach—to kill pathogens and then discharged into Upper New York Harbor.

The quality of this effluent is open to interpretation; it meets federal and state standards for fecal coliform (bacteria that inhabit the intestinal tracts of mammals) and other pathogens, but chlorination does nothing to break down hormones and antibiotics in the waste stream. Doctors routinely advise patients to flush unused or expired prescription pills down the toilet, and hospitals do the same with drugs unused by discharged patients. But scientists have become increasingly alarmed about the effect of excreted pharmaceuticals, including birth control pills, steroids, antibiotics, pain pills, and Prozac, on wildlife in and near our nation's waterways. Within the last decade or so, endocrinologists have correlated deformities and behavioral changes in fish, amphibians, and birds to low, but constant, levels of endocrine disruptors that flow, with urine, into and then out of wastewater treatment plants.

Leaving the rectangular tanks, Heckler and I ducked into a windowless building filled with four open cone-shaped thickening tanks, sixty feet across. Huge feed lines and ducts led in every direction. Skylights dimly illuminated the passageways. There wasn't an employee in sight: the whole plant ran, on rain-free weekends, with a crew of just seven. It was spacious and warm in the tank building, and a quiet pumping sound filled the air. If you could forget for a moment that you were in a concrete shithouse, the place seemed almost tranquil.

We climbed to the building's highest level, where I ambled out onto a metal catwalk over a thickening tank. Below me, a ring floating atop the swirling brown batter reduced turbulence while rotating scrapers pushed liquid to the edge and solids to the bottom. Heckler called this mixture, which was 97 percent water, "dissolved solids." I noticed an orange life-saving ring hanging on a nearby hook and experienced an unpleasant moment of vertigo. The catwalk was awfully narrow, and the guardrail mighty thin.

After two hours of thickening, which reduced water content by 1 percent, the sludge was pumped to six thirty-eight-foot-deep tanks called digesters, which worked with centrifugal force. Here, as in the aeration tanks, old sludge was mixed with new as a microbial starter. The bugs in the digester, much like those in a human stomach, preferred a temperature of 95 degrees. Decomposition at this stage was anaerobic, and it produced the methane that ran the engine I'd seen at the start of my tour. The engine drove a generator that produced electricity for the plant.

Heckler and I wandered around the digester building, up metal stairs and down, around concrete columns and under huge ducts, trying to locate the exit. It was like that scene in *This Is Spinal Tap* where the band can't find the stage. As we walked, Heckler explained the end game. The sludge continued to cook in the digester for fifteen days, then it was pumped toward the shore, where a tanker pulled up once a day. The boat, which held ninety thousand cubic feet of finished product, pumped its load into a holding tank at a dewatering plant on the East River. After spinning around a centrifuge and accepting a dose of polymers, which encouraged clumping, the sludge went up to 26.5 percent solid. In its next move, the sludge would exit the province of the public and enter that of the private. Soon it would have some value.

Eventually Heckler and I found a door and emerged into bright sunlight on the north side of the plant. We were on Harry Smutko Walk, according to a street sign. Heckler didn't know who Smutko was, but later I called the plant manager, Bill Grandner, who told me that Smutko had worked at the plant in the late sixties. "Unfortunately, he died in the facility," he said. "He either fell or he tripped, after a heart attack—forensics weren't so good back then." What did he fall into? I asked. There was a slight pause before Grandner answered: "It was a final tank." Indeed.

Chapter Eleven

In the Realm of Taboo

Owls Head was the end of the line for the water I flushed from my apartment, but the solid portion of my effluent had miles to go. Somewhere between the treatment plant in Bay Ridge and a factory on the South Bronx waterfront, my sewage was transformed, semantically, into "biosolids." The neologism had been forged in the crucible of public relations and fired by the potential for profit.

For decades, the DEP had dumped twelve hundred tons of sewage a day from a tanker parked twelve miles off the city's shore. But in 1985, the EPA pronounced the waters near the Twelve-Mile Site officially dead. The only shellfish that remained were contaminated with bacteria and heavy metals. Fish showed accumulations of metals and toxic chemicals. In 1988, the city switched to a new site, 106 nautical miles southeast of the harbor. The sludge tankers, which couldn't make the hundred-hour trip, were converted into pumper vessels that filled newly built long-haul sludge barges. Before long, though, commercial fishermen who worked near the 106-Mile Site began to complain of decreased catches and sick fish. Other old and densely populated cities have had similar problems disposing of their sludge. Starting in 1878, Boston's sewage was held on Moon Island: on the outgoing tides, the facility's gates swung open and the untreated waste was flushed into Boston Harbor. Congress made this nasty habit illegal in 1988 with the Ocean Dumping Reform Act. Boston waited until 1991, when the ban went into effect, to quit dumping its 400,000 daily gallons of sludge into the ocean; New York's last load chugged out of New York Harbor, aboard the *Spring Creek*, on June 30, 1992.

Where would all that sludge go now? Onto farm fields, said the EPA, which had recently reclassified the material, after it was treated to reduce pathogens, from a hazardous waste to a "Class A" fertilizer. The new rules governing sludge policy, dubbed Part 503 by the EPA in 1993, raised the acceptable exposure limits to such toxins as lead, arsenic, mercury, and chromium so that most of the nation's sludge could be classified as "clean." With this regulatory makeover, a new era of "beneficial use" for sludge began. But first, the product needed a new name. Who in their right mind wanted to spread municipal sludge around his backyard? The sewage industry's main trade and lobbying group, known today as the Water Environment Federation (WEF), decided to sponsor a naming contest. (The group also dreamed up the Select Society of Sanitary Sludge Shovelers, which honors workers who go above and beyond the call of duty. Members, who wear tiny silver shovels on their left breast pockets, have a special handshake and a password derived from the first letters of the society's name, pronounced "Sh-h-h-h.") According to John Stauber and Sheldon Rampton in their excellent book *Toxic Sludge Is Good for You!*, WEF members made more than 250 name suggestions for the contest, including "purenutri," "bioslurp," "black gold," "geoslime," "sca-doo," "the end product," "humanure," and "hu-doo." In 1991, the name change task force settled on the comparatively bland "biosolids."

By then, many cities were already branding their homegrown

product: Chicago sold Nu-Earth, Los Angeles produced Nitrohumus, and from Houston came Hou-actinite. Milorganite, which Milwaukee had been selling since long before the regulatory renovation, had its own advertising campaign, which included broadcasts at Brewers games. "Give your grass the all-you-can-eat buffet it craves this time of year," an announcer sang out. "Milorganite organic nitrogen fertilizer . . . is packed with the nutrient home-run power that grass loves." (Milorganite buyers were warned against applying the stuff to food-producing soil; in 1982, Maryland scientists found such high levels of cadmium in Milorganite that they banned its sale in their state.) Today, 54 percent of the 7.5 million dry metric tons of sewage sludge that the country disposes of each year is processed, relabeled "biosolids," and applied to land. The rest is buried in landfills (28 percent), incinerated (17 percent), and "surface disposed" without processing (1 percent).

New York City's own sludge had no advertising budget: in fact, it was hardly known within the state's boundaries, though it was processed into tiny balls of fertilizer called Granulite right inside the city limits, in the Hunts Point neighborhood of the South Bronx, by a company known as NYOFCo, the New York Organic Fertilizer Company. I headed uptown one morning, eager to see how this little-known operation handled more sewage than any other pelletization plant in the nation.

Ten thousand people live in Hunts Point, though it is zoned primarily for industry. The neighborhood includes two-dozen waste transfer stations, a raw-sewage dewatering facility, the sewage sludge processing plant that I was about to visit, and the largest produce distribution center in the world. Together, these industries generate more than twenty thousand diesel truck trips per week. In 1980, Hunts Point was identified as the southern tier of the poorest congressional district in the country, and in the 1990s, the neighborhood became even poorer. More than two-thirds of its residents under the age of eighteen live below poverty level. The majority of residents are Latino and black: their asthma rate is the highest in the country.

Early one winter morning, I buzzed at NYOFCo's security gate and a scarf-swaddled employee on a small plowing tractor waved me into the snowy parking lot. I noted the air's loamy smell, and then Peter Scorziello appeared from out of nowhere, greeting me in his shirtsleeves in 20-degree weather. He opened the door to a deserted lobby, and I took my first deep breath of what amounted to the ultraconcentrated back end of New York City. The smell wasn't loamy anymore. "I think of it as a musty, cheesy odor," Scorziello said.

I considered the comparison for a moment, then rejected it. This wasn't the aroma of any cheese I'd ever sniffed. Nor was it the ammoniac smell of a port-o-san or the earthier tones of the outhouse. It wasn't the chicken or cow manure of my childhood garden. Elusive and remote, the name of the scent lingered just out of my reach (unlike the odor itself, which would stay with me for several hours to come).

The second-floor conference room was dominated by a gleaming wood table ringed by metal chairs. In the elevator going up, I'd hoped it would offer some respite from the smell, but no. The odor might even have been worse in here, but Scorziello didn't comment on it so neither did I. My host was a mild-mannered and down-to-earth chemical engineer with a hesitant beard. He had picked the job with Synagro, NYOFCo's parent company, from the newspaper's employment listings just a few months after graduating from college. "It was a job; it paid the mortgage," he said with a shrug. He started out as a shift supervisor and now, after ten years, managed the plant and its thirty-one employees.

Before New York City had contracted with NYOFCo to pelletize its biosolids, the company trucked untreated sludge, which is known as "cake" and classified as Class B fertilizer, seventeen hundred miles to Colorado, where it was distributed on rangeland. (Class B sludge has a higher pathogen level than Class A and cannot be applied to land that grows food for human consumption.) "It's just thousands of square miles and one guy spreading it,"

Scorziello said, a faraway look in his eyes. I doubted Scorziello was nostalgic for those days, as pastoral as they sounded. Spreading New York sludge on open ranchland left a bad taste in many small-town mouths. When New York quit ocean dumping in 1992, it turned to a company called Merco Joint Venture to handle its Vesuviuses of waste. But the company immediately ran into problems. When it tried to deliver its filled-to-the-gills railcars to Oklahoma, the state banned disposal of out-of-state sludge. When it tried Arizona, the state blocked rail shipments. Finally, after Merco made a donation to Texas Tech University to study the beneficial uses of sludge, a deal to dump in the tiny southwestern town of Sierra Blanca fell into place. According to a Texas Water Commission official quoted in the *Dallas Morning News,* "This thing was pushed to the top of the stack. Giving a $1.5 million grant to Texas Tech helped." The company never performed an environmental impact statement, nor did it solicit citizen input.

Years ago I visited this desperately poor town, which lies eighty-eight miles southeast of El Paso and was home to just 650 people, about 40 percent of whom were poor. I was in Texas to report a story about the so-called cleanest town in the country, which happened to be just about an hour away (if you drove like a Texan). While "chemically sensitive" people, intolerant to minute amounts of chemical pollutants, flocked to Fort Davis to breathe the pristine mountain air (in homes they wallpaper with tinfoil, to block the possible outgassing of arsenic wood preservatives), residents of Sierra Blanca sucked the rank dust that wafted off Merco's 81,000-acre dump site. Three times a week, fifty flatcars loaded with minimally treated cake rolled up to the Merco property—usually at night. "It smelled like death with a chemical odor," said Bill Addington, a local who fought the sludge farm. "A state highway ran through the property, but I'd drive fifty miles out of my way to avoid it because my son threw up when we went through." Others blamed the sludge farm, the largest in the world, for their rashes and mouth blisters, their asthma and increased allergies, flus, and colds. While I was visiting, less than a year after the start of dumping, an entire colony of Mexican free-tailed bats, which had roosted under a nearby railroad trestle since the 1880s, was in the midst of annihilation, victims of chemical poisoning.

In August of 1994, EPA tests of Sierra Blanca sludge showed levels of fecal coliform at thirty-five times the acceptable rate. Merco was fined $12,800, and the poo-poo train, as residents called it, kept chugging. Disgusted and sickened, opponents in 1997 filed a civil rights complaint with the EPA against the Texas Natural Resources Conservation Commission. The complaint was denied, and Merco won a five-year renewal of its sludge permit—a contract worth $168 million—and permission to up its daily dump from 250 tons to 400. In 1999, the company was again caught violating federal and state regulations by improperly treating sludge for bacteria and pathogens; this time it was advised to mix the cake with lime. Finally, nearly a decade after New York's sludge had landed in their town, Sierra Blancans got some relief. New York City decided that shipping biosolids 2,065 miles was no longer cost-effective, and it canceled its contract with Merco. Almost immediately, the company declared bankruptcy. The next company to handle the city's sludge would be Scorziello's.

In the NYOFCo conference room, we stood before an easel that held a large schematic of the plant. "In May of 1993," Scorziello said, "we converted from Class B fertilizer to pellets. Now we take in about 575 wet tons a day and send out 138 dry tons, for a total of 210,000 wet tons a year." Accumulated over the course of a year, the pellets would fill 503 railcars.

"There are very strict rules on how and when sludge can be applied," Scorziello continued. NYOFCo adhered to those rules through a process of drying and heating. (The regs could also be met by stabilizing the biosolids with lime or by composting them for forty days at high temperature.) The process started when Scorziello's truck drivers collected the day's raw material from nearby dewatering plants. On the easel, I traced the biosolid's path through the plant with my finger: from the receiving floor to a mechanical screw and onto belt conveyors. "It's like a moist cake at this stage," Scorziello said. From there, the biosolids fell into a storage hopper and then a mixer. "Now they have the consistency of tapioca."

Gravity pulled the biosolids into a dryer, where they tumbled around a drum blasted by 600-degree air for thirty to sixty minutes. The pellets then went into a separator can, where a screen sorted them into four sizes. The stuff that was slightly too big ("nonmarketable nuggets," Scorziello called them) went to a crusher. The stuff that was really too big went to the dump. I asked if these chunks had to go to a specially regulated landfill. "No, they're not hazardous or toxic," Scorziello said. "They could be used as beneficial cover, except that then they'd have to pay *me*." He chuckled. (In fact, sludge sold as fertilizer can be so contaminated with toxins that it can't legally, under the EPA's Part 503 rules, be buried in landfills designed for household waste.)

Pellets that were too small were recycled within the system, while platonically ideal pellets whooshed pneumatically through pipes into eight storage silos. Twice a day, a railcar pulled forward on the plant's eastern side and took on a hundred tons of Granulite. Most of the trains went north toward Albany, crossed the river at Selkirk, and then turned left for Florida, where the sewage of New York City was spread on citrus groves. Other trains made their way to Ohio, to corn and soybean fields.

I asked Scorziello why Granulite wasn't used in New York. "I think it's too easy to apply, so road crews working on highway medians don't get overtime. Spreading composted biosolids takes longer." (There were other, unstated, reasons the pellets didn't stay local. Sludge is produced every day, but fertilizer can't be spread year-round in the Northeast because the ground freezes. NYOFCo, which produced an enormous amount of product, wanted a steady taker. Moreover, northeastern soils tend to have a low pH, and many scientists believe the EPA 503 regulations don't adequately protect alkaline soils from the leaching of metals.)

I picked up a plastic bag from a window counter. It weighed about a pound and held pellets the size of peppercorns. They looked hard and dry. "Smell it," Scorziello said. Like a pot smoker sniffing new product, I took a quick whiff. The words *rich* and *loamy* came to mind, but so did *manure*. I shook some Granulite onto my hand, just to see what holding someone else's highly processed feces felt like. It was no worse than handling raw meat, in the sense that it was so recently part of a living organism.

Scorziello handed me a hard hat, goggles, and a gauzy white smock. "You can leave your jacket here," he said, then added, "Hmmm. Better leave your fleece vest, too." We left the conference room through a back door that led to a catwalk overlooking the production floor. It was a little noisy out here but not too bad, with vents and shafts running all over the place. The air smelled strongly of ammonia. We walked downstairs and along a conveyor belt dotted with round elephant piles of dark gray paste. This raw material, as Scorziello called it, was about 74 percent water at this stage. Its consistency was strangely reminiscent of the toxic mudflats that I'd recently paddled around at Fresh Kills. When Scorziello was done with this material, its water content would be down to 4 percent.

Sticking a hollow metal tube into a machine, Scorziello pulled out a sample and passed it from hand to hand. The look on my face prompted him to say, defensively, "It's not feces." I supposed, with bacterial pathogens missing, that it wasn't, but that didn't change my gut-level reaction.

Scorziello brushed his hands lightly against his khakis, and we climbed up to the control room, where two workers monitored two thousand different manufacturing parameters on a bank of computer screens. We could have continued walking around the production floor or looked out the windows that surrounded this booth, but instead Scorziello offered me a virtual tour of the six identical "trains" that tumbled, dried, and separated the biosolids. Afterward, we clomped back downstairs to the receiving area. "Smells like something's burning," I said.

"That's the reject material," Scorziello said. A tall pile of dung was steaming away on the tipping floor. Stored for about a month until there was enough to dump, the oversize biosolids got hotter and hotter. The bay doors were open, and I asked about vector control. The EPA 503 regs are quite strict about controlling animals that might transmit diseases from biosolids to humans. He said there were rats in the stuff now and then. And raccoons wandered in the open doors. In the summer there were flies. "It's annoying, but it's not swarms," he said. There were stray dogs in the neighborhood as well. A truck driver had adopted two and named them Pellet and Sludge.

Scorziello was a scholar of smell. The company sent him to odor seminars, and odor engineers made monthly surveys of his factory. Before air was released from the plant, it whooshed through a Venturi scrubber that removed most of the fine dust and ammonia, then through a regenerative thermal oxidizer set for 1,620 degrees Fahrenheit. A continuous monitoring system on the stack told Scorziello what he was discharging at any given moment.

"In 1996, we had some issues," Scorziello said. "We got all the engineers together and found a problem in the regenerative thermal oxidizer." To understand the smell, it was necessary to parse it. "The fecal odor," Scorziello explained, "is a mixture of skatole, valeric acid, and butyric acid." I knew that valeric acid makes rotten-cheese smell, and butyric acid is the smell of vomit. Later, I looked up skatole and learned that it's an organic compound found naturally in feces and beets. Because it prevents rapid evaporation, it is also used as a fixative in the manufacture of perfume.

"Each of those compounds has a high boiling point," Scorziello continued. "They were condensing in our oxidizer and dripping out onto the floor." With some tweaking, Scorziello lost the condensation. The company recently spent at least another $2 million on equipment changes, but some of the neighbors remained unimpressed.

Omar Freilla, program director of Sustainable South Bronx, which has been fighting NYOFCo for three years, thought the plant smelled "horrible." "It's beyond an ordinary sewage smell," he said. "It's in between sewage and fish, with a chemical odor thrown on top. It stinks from the inside. When I got back from a tour of the plant, people would not stand next to me, they sprayed me with Febreze. The smell comes with nausea and headaches and tightness of breath, if you're asthmatic."

"I've seen children throwing up from the smell," said Eva Sanjuro, who runs a day-care center a few blocks from the plant. "Teachers have quit because of their asthma. When it's bad, we don't bring the kids outside. We call the plant, and they blame the water treatment plant or the chicken butcher." When the DEP received complaints, the agency sent an inspector. But he or she never arrived in time, said Freilla, to finger the culprit.

Though warned against flushing anything but the basic materials down their toilets and drains, New Yorkers, like people all across the nation, routinely pour out bleach, paint, and nail polish remover, among other household toxics. Though industry was required to pretreat its effluent in New York City, the DEP annually cites hundreds of radiator repair shops, drum refinishers, paint manufacturers, metal platers, and circuit board manufacturers for dumping chemicals. Thirty years ago, more than 9,000 pounds of heavy metals entered the sewer system each day; today it's down to about 2,900 pounds. Some of those toxins are neutralized in the wastewater treatment process, but treatment plants were designed with the Clean Water Act in mind: they preferentially remove contaminants from water only to concentrate them in sludge.

New York's sludge contains the metals magnesium, cadmium, zinc, iron, mercury, selenium, lead, and copper. "Plants need copper, zinc, and calcium," Scorziello told me. "It makes very green foliage. It increases drought resistance and helps soil retain water. At these levels, the metals are micronutrients." According to the Government Accounting Office, sewage sludge nationwide also contains PCBs, pesticides, asbestos, DDT, and dioxin.

An EPA presentation on dioxin to a committee of the National Academies' Institute of Medicine, in April of 2002, indicated that land-applied sewage sludge is the second-largest source of dioxins in the United States, second only to backyard barrel burning. A Natural Resources Defense Council review of the scientific literature confirmed that dioxin is taken up by plants that are grown on sludge and is stored in the fat tissue of animals that graze on them. The environmental group, in 2003, unsuccessfully sued the EPA to limit dioxin in sludge. Perhaps not surprisingly, considering the influence of private industry in this country, the United States is markedly less stringent about regulating sludge than other countries.

Dioxin isn't the only thing to be worried about. A large proportion of pollutants are assumed to bind tightly to soil particles, making them inaccessible to organisms feeding there. But in 2002, scientists from the University of Franche-Comté in Besançon, France, placed snails in cadmium-laced soil collected near a smelter. After two weeks, the scientists analyzed the snails' tissue and found that about 16 percent of the cadmium they had absorbed was supposedly inaccessible. The finding suggests that other heavy metals may be more "bioavailable" than assumed and could be entering the food chain. According to the World Health Organization, long-term exposure to cadmium, partly from fertilizer and food, leads to kidney damage, cancers, and possibly birth defects. Other industrial by-products that end up in sludge have been known to cause neurological, immunological, and other problems in people and animals.

I asked Scorziello about hospital waste flushing through the city's system. I was thinking about unused liquids from radiotherapy or laboratory research, and urine and excreta from patients treated or tested with radionuclides, which are used for in vivo diagnosis. "Our stuff could be radioactive," Scorziello said casually. "But we don't test for it." In 1999, the city of Denver developed a plan to pump plutonium-contaminated water from a landfill—a Superfund site—into the municipal wastewater system. The treatment plant produced a "beneficial biosolid," marketed as Metrogro, that was spread on farmland. A sewage agency board member blew the whistle on the scheme, but the sludge—with radioactivity levels higher than federal standards but lower than state—continues to be spread on fields that grow wheat that is sold to granaries worldwide. Though some plants can and do take up some radioactive isotopes, data from wheat grain tests performed by the US Geological Survey were "insufficient to determine any measurable effects from biosolids."

The EPA didn't require NYOFCo to test for dioxin, PCBs, or radioactivity, but it did require readings on lead, which leached from household pipes and ended up in sewage. The allowable level for Class A biosolids was fewer than 300 parts per million. "We're usually at 150 to 200," said Scorziello. "If I get a load with lead that's too high, like 305, it's okay. The average is over a month. I just tell the grower, and they know where to apply it." I had a funny feeling about that. Farmers don't apply an average when the truck pulls up: they apply the load that arrives in their driveway. And it was hard to believe that the laborers hired to spread those pelletized biosolids were keeping close track of where and when they spread each delivery of fertilizer. When the EPA assessed vegetables grown in soil amended with biosolids, it found no significant health effects from eating these vegetables when the trace metals in the biosolids had been applied at regulated rates. But who did the regulating? And how could a grower regulate if she didn't know what metals, to say nothing of their levels, the fertilizer contained? NYOFCo's Granulite label lists its nitrogen, phosphate, and soluble potash content, but not its lead or other heavy metals, their amounts and possible interactions. It warns users only to wear gloves when handling the product and to avoid ingestion and inhalation.

Of course, there isn't anything intrinsically wrong with returning nutrients from organic matter to the soil. Human and animal waste contains high levels of nitrogen and phosphorous—stuff that makes plants grow. Ancient cultures have long fertilized their farm fields and vegetable gardens with human and animal manure. But Granulite's packaging describes the product as an "all natural organic" fertilizer. The words *sewage sludge* appear nowhere on the label. "Clearly inputs to sewage treatment plants in New York City, or anywhere, are not all natural and organic," said Ellen Harrison, director of Cornell University's Waste Management Institute, which has been critical of the EPA's sludge regulations. "That qualifies as false advertising in my mind."

With its very name, the New York Organic Fertilizer Company trades on the feel-good ethos of composting, of enhancing soil nature's way. But in fact, federal law prohibits growers from using NYOFCo's product, or any sewage product, on organic fields. It doesn't matter if the municipality combines household and industrial wastes or not. "You just don't know what's in the sludge," said Carol King of the Northeast Organic Farming Association.

In 1999, Representative José Serrano, of the Bronx, introduced a bill that would require food products grown on sludge—including crops and livestock or poultry raised on land treated with biosolids—to be labeled as such. The EPA strongly denounced the bill, fearing it would spark unwarranted fears; the bill remains in limbo. Now and then, activists cry for honest labeling of biosolids; Milwaukee, whose wastewater treatment bureau has a million-dollar PR budget, was rumored to have threatened to sue the EPA if it required detailed labeling of Milorganite.

There have been threatened lawsuits from the other side, too. Across the country, people claim to have been sickened by sludge. Anecdotal evidence links Class B biosolids with abscesses, reproductive complications, cysts, tumors, asthma, skin lesions, gastrointestinal problems, headaches, nosebleeds, and irritation of the skin and respiratory tract that leave people vulnerable to infection. After Shayne Connor, a twenty-six-year-old New Hampshire resident, died in 1995 from a staph infection following the application of biosolids on a nearby field, his family sued Synagro, which owned the company that spread the sludge, for wrongful death. The company settled out of court for an undisclosed sum. Parents of two children in Pennsylvania also alleged, in separate incidents, that the children had died of staph infections, in 1994 and 1995, after sludge was spread near their homes.

Farmers in Georgia are suing municipalities for improperly treating the sludge applied to hay fields: their cows became sick and died after eating the hay. A California town was cited for excessive sludge applications after thirteen cows on two farms died of nitrate poisoning. The Cornell Waste Management Institute alone has received more than 350 sludge-exposure complaints from communities across the country, and thousands of people have reported sludge-related health problems.

In 2003, seventy-three labor, environmental, and farm groups called for a ban on land application of biosolids. In January of 2004, the EPA denied their petition but announced it would reassess fifteen additional inorganic hazardous chemicals found in sludge, beyond the nine pollutants it currently regulates. If the studies warranted, the agency said, new regulations would be proposed.

Sludge activists don't put much faith in EPA data: the scientific studies on which the agency based its original sludge regulations were performed by the sludge industry itself. And to investigate nineteen complaints of alleged sludge-related illnesses or property damage, the EPA gave grant money to the sewage industry's main lobbying group, the Water Environment Federation. Within the last few years, the EPA inspector general, the Centers for Disease Control and Prevention, and the National Research Council have all issued reports urging more research into the effects of sludge. Yet in May of 2003, the EPA microbiologist David Lewis, who had challenged sludge safety in the esteemed journal *Nature*, left the agency under cloudy circumstances. Lewis claims he was fired for questioning sludge standards; the EPA claims Lewis signed an agreement specifying that he would step down.

One of NYOFCo's biggest customers was Lykes Brothers Citrus, in Lake Placid, Florida. The Lykeses used to be Tampa's richest family. They got their start in business shipping cattle to Cuba during the Civil War, and in the early 1900s branched out into ranching in Texas. At one time, the family ran the largest US-flagged cargo fleet in the nation. The Lykeses were involved in overseas meatpacking operations, real estate, logging, banking, oil and gas, the insurance business, and a trucking company. Though the family had squabbled and divested in recent years, the Florida Lykeses still owned a fertilizer company, and they grew pine, eucalyptus, sugarcane, and citrus, and raised cattle. By spreading sludge instead of fancy fertilizer, growers like Lykes saved about $5,600 for every hundred acres.

My contact person at Lykes was Bill Barber, but our contact was brief. He said nothing when I told him my biosolids were feeding his fruit. When I asked how much of the stuff Lykes used, he announced, "We have a company policy not to talk about things like that." I asked if he would at least confirm that Lykes used NYOFCo's product. "I'm sorry, I'm not going to talk about that. We appreciate your call. Thank you." Then he hung up.

It's natural to wonder why all biosolids aren't groomed for Class A status: Class A fertilizer seems a lot safer than Class B. But because Class B is less processed, it makes a far better fertilizer than Class A: it has a higher nitrogen value. There are tradeoffs: Class B is cheaper to produce, and you get more of it from a load of sludge than you do making Class A, but Class B is wetter and heavier, and therefore more expensive to transport. Class B also smells worse than Class A, which compounds its PR problems.

Scorziello was about as proud of his product as he could be. Because the pellets were forged in high temperatures, he told me, "it's almost impossible to have nitrogen burn with them. They release their nutrients slowly." I asked him what this wondrous material sold for, and when he gave me the answer my jaw dropped. NYOFCo gave their biosolids away for free! Not only wasn't Bill Barber paying for his pellets, he didn't pay for their delivery, either. NYOFCo picked up the shipping tab—between $15 and $45 a ton. "Yep," Scorziello said. "We make our money from the city. They give us $131 a ton to process it. We're a residuals management company," he said. "We're not a fertilizer company. We perform a service for the city."

With that in mind, my entire perspective changed. NYOFCo wasn't here to produce a valuable product and sell it to growers; the company's mission was to take something that no one else wanted and move it somewhere else with minimal hassle. Such was the refrain of the waste industry.

It was early afternoon when we finished talking, and Scorziello offered to drive me to the train station. "Thanks," I said. "I'm just going to wash up before we go." "Good idea," he said. "I will, too." There was a shower stall in the women's bathroom, and an enormous jug of antimicrobial soap.

Scorziello drove a new Volvo SUV, which had a small black gadget called an Ionic Breeze attached to the console. "My wife got it for me from the Sharper Image," Scorziello said. The deodorizer trapped airborne particulates on an electrostatically charged ring. I supposed it was working, because the car still smelled pretty new (that is, like off-gassing PVC) despite having hauled its share of Granulite back home to Tarrytown.

"I use it every holiday on my tomatoes and flower garden," he said. "You know, Easter, Fourth of July, Thanksgiving, and Memorial Day. My neighbor hangs over the fence and says, 'You using that stuff again?' But the smell dissipates in a day." Until recently New York State had prohibited the use of sludge or sludge products on crops for direct human consumption; Scorziello had never paid the regulation any mind.

"Do you compost?" I asked just before we said good-bye.

"My wife won't let me. Also, the development doesn't allow it."

"Why not?"

"I'm not sure. There are probably vector control issues. And the development can't control what people put into their compost." Ditto with the city and its sludge, I thought.

I said good-bye to Scorziello and got onto the next train. Within moments of sitting down, though, I realized that something was terribly wrong. With my every movement, my down jacket released a puff of eye-wateringly rank air. People were actually moving away from me. After riding three trains, the longest commute of my life, I arrived in Park Slope. My husband got a whiff of me and pulled a bad face. "I'm taking a shower and washing all these clothes," I said.

"Good idea," he said, and quickly closed his office door.

I was lucky to live in a neighborhood that flushed and forgot its sewage. I didn't live within smelling range of the Gowanus Canal, the Owls Head treatment plant, or the pelletization plant in the Bronx. I'd probably never breathe the noxious dust of New York's pelletized excrement in Ohio or Florida. That I had only a vague notion of my waste's impact on others troubled me. It was easy to shrug off those effects if they weren't harming you personally.

What could we do better? Arcata, in northern California, opted out of building an expensive sewage treatment plant and spreading its sludge on farm fields. Instead, it constructed an artificial wetland within the city limits. Untreated waste flows through a series of plant-and-animal-filled ponds and comes out the other end purified. The marsh is a beauty spot, where locals come to jog or to sit on benches and watch more than 160 species of birds. The city of South Burlington, Vermont, did something similar, in a pilot project, inside a greenhouse.

Wetlands probably aren't practical for most big cities. We have a lot more sewage to treat than Arcata, we've already filled in most of our natural wetlands (with garbage), and it is unlikely that the city or private landholders will give up space to create sewage-treating marshes. But there is one place in New York where space is seriously underutilized: our rooftops. We have thousands of acres of them! Several city groups are working to establish green roofs atop residential, commercial, and government buildings. The advantages of green roofs are numerous: they keep buildings cool in the summer and warm in the winter, they clean the air, they mitigate the city's heat-island effect, and they absorb storm water and slowly release it into the atmosphere. This is no minor point. Scientists have recently warned that the rising global climate will soon bring more storms and higher sea levels to New York City, both of which will threaten our already crumbling infrastructure, including our wastewater treatment plants. By 2100, say climate specialists at the Columbia Earth Institute, the probability of a one-hundred-year storm (one so drastic that it occurs, on average, only once in a century) could be, in a worst-case scenario, one in every four years. A 14.5-foot storm surge produced by such an event would bring the Gowanus Canal within a few avenues of my house.

The more storm runoff we can keep out of the sewer system, the better. But some visionaries take this a step further: they imagine miniature rooftop wetlands that also handle the building's gray water and process its sewage into food for plants. It sounded utopian to me; I liked the idea. But it wasn't coming to my building soon. (Why not? Because it would void my roof warranty, and it would take far too long to recoup my investment.) Rooftop wetlands aside, is there any good place to put ever-increasing mountains of sewage sludge? In developing countries, sludge is sometimes fired to form bricks or composted for farm fields. But developing nations have only a fraction of our problems with contaminants. I've always been fond of the Clivus Multrum, a waterless composting toilet in which solid waste, mixed with sawdust, is reduced by microorganisms to 10 percent of its original volume. A tiny battery- or solar-powered fan directs any bad smells out a roof stack. The end product is an odorless humus, perfectly safe to use on home gardens.

I wasn't quite ready to dismantle my toilet, though thousands are moving in this direction. More and more people, living off the grid or not, have recognized how little sense it makes, when our population is so large and our clean water supply shrinking, to dilute our solids with water and then, at great expense, separate the two. The more I learned about the potential value of things that we call waste—and the more I was dealing with my waste instead of sending it out of sight—the more blurred the lines became, in my mind, between sewage and garbage, sewage and compost, and garbage and compost. Once again, waste was looking a lot like food.

I heard about a graduate student in agriculture at the University of California, Davis, who had built himself a home gray water treatment system. Perhaps the most important prerequisite for such a labor-intensive undertaking, I learned when I phoned Tim Krupnik, was having a laid-back landlord. Krupnik was living in a group house in North Oakland when he basically "took a Sawzall trip under the house." Cutting through Sheetrock and beams, he unhooked the shower drain and diverted it to his backyard. "Our model was a natural wetlands," he said. "The water flowed into a pit, a little wider and deeper than a bathtub, lined with gravel of different sizes."

Krupnik first planted the pit with cattails that he'd collected along with spoonfuls of microorganism-rich mud from San Francisco Bay. The plants got their phosphorous from Krupnik's biodegradable soap and shampoo and their nitrogen from skin and hair in the runoff. Then he installed more aquatic plants, including taro. "I know it's nonnative, but I wanted to see how it would do." It did well, reaching six feet tall.

From the tub, Krupnik's gray water spilled into a secondary cleaning system, a concrete basin set into the ground. "I put some water hyacinth in there—they have a complex fractal root system, a lot of surface area, space for microorganisms to live and reproduce." He hooked a pump to his solar panel and aerated the basin. Then he added goldfish, to eat the mosquitoes that were soon breeding there. The fish did well for months and months without the addition of fish food. "But we had eight or nine people in the house, and if someone took a very hot or very cold shower . . ." Krupnik's voice trailed off, but I got the picture: fish are temperature sensitive. He used his gray water to irrigate the yard and fruit trees, but not vegetables: "If there are pathogens in the water, they'd concentrate in the stalks."

"I heard that you're recycling more than shower water."

There was a pause on the other end of the line. Krupnik knew what I was talking about. "Ah," he said. "Now you're getting into the realm of taboo."

After setting up his gray water system, Krupnik decided to go a little further. He built a table about the size of a nightstand. The top opened to a toilet seat; underneath was a two-and-a-half-gallon bucket. "It was only for feces," he said. "And I added a cup of sawdust each time to keep the smell down and soak up water."

Every few days, when the bucket was full, Krupnik carried his humanure—that was the word he'd decided on—outside to compost it with red worms in a bin. "The worms eat half their weight every day," he said, echoing the enthusiasms of Christina Datz-Romero, of the Lower East Side Ecology Center. "It's your typical thermophilic compost. The idea is to let it reach 120 to 160 degrees for several days. That kills the pathogens." To conserve moisture, keep out flies, and add carbon, he worked strips of newspaper into the surface. "I had a two-to-one carbon-to-nitrogen ratio," Krupnik said. In less than two months the bin was filled with high-quality compost.

Krupnik kept his toilet in his bedroom. "People would come over to watch a video and they'd put chips and salsa on the nightstand," he said. "They'd sit on the couch and reach over. They never knew what was going on. I'm a vegetarian, and because I monitor my carbon-nitrogen ratios closely and keep a lid on the bin, I've never had any problem with odor."

He used his finished product on ornamental plants and fruit trees. His red apples, he said, were delicious. "Trees protect their seeds and fruit. I wouldn't use the compost on something like lettuce, though. It's a lot of biomass to go into that kind of plant." He warned against composting excrement lightly. "You need to do it with care," he said. "There could be E. coli in it if you're sick. You need to read up on it, watch what you eat." He suggested *The Humanure Handbook,* written by a former roommate's father, Joseph C. Jenkins. "That's undoubtedly the best resource."

I looked up the book on Amazon. Thirty-eight customers had reviewed it, giving it an average rating of five stars. "The best book since the Bible," wrote one reader. "I can't wait to build my own sawdust toilet and finally add my own poop to nature's cycle, the way I've wanted to since I was six years old!" wrote another. Someone else wrote, "I found the book to be a life changing force, in that you just cannot flush the toilet anymore without thinking about all the wasted water or toxic chemicals going into that act."

At the 2003 World Water Forum in Kyoto, where government ministers and water experts discussed ways to meet a United Nations target of halving by 2015 the number of people in the world who lack clean drinking water and modern sanitation, scientists recommended that communities in developing nations compost their biosolids rather than invest in sewer hookups, which would empty into rivers and create public health disasters. At the meeting, the United Kingdom's former "chief drinking-water inspector" said that if Britain were planning sewage disposal from scratch today, "we wouldn't flush it away—we would collect the solids and compost it."

Composting human excrement, I was coming to realize, was far more than a gardening issue: for many, it was political as well as environmental. There was a growing constituency out there that wanted nothing to do with what they considered a wrong-headed system, in which clean water was sullied with excrement, the valuable nutrients in human waste were combined with industrial toxins, and municipalities picked up the tab for disposing of industry's mess. Raw sewage—a biological nutrient—belonged to the people, in this view.

I asked Krupnik what had inspired him to make humanure. "I wanted to see if I could do it," he said. "And because I've been in recycling for a number of years, I wanted to practice what I preach. I get a big sense of personal gratification from being able to connect myself to a system. You eat the apple, you metabolize it, it goes into the toilet, you compost it, and you put it back on the apple."

I identified completely with that sense of connection—I felt it every time I put some carrot tops in the compost pile and imagined that Lori and Simon, who lived in our building's garden apartment, would grow something nice to eat with it. Krupnik continued: "I like being part of a productive system. I like being able to live in Oakland as if I'm living in the mountains." But people living in the mountains, I pointed out, probably spent a lot less time worrying about their environmental impact: lower population density meant they had less impact, and the impact probably wasn't right under their nose. I wondered if Krupnik's fussing with his shit was yet another example of self-indulgence, the luxury of a guilty urbanite with sufficient time to ponder his outsize environmental footprint. But at least he was doing something about it.

* * *

Garbage and excrement are linked on the dark side of human nature. Freud suggested we are drawn to touching excrement precisely because it is taboo. Waste products are primal, and although they clearly stand for the back end, for death, they are also linked inextricably to life. Animal waste feeds new plants and new animals. Nutrients from the dead, abetted by fungi, bacteria, and other agents of decomposition, jump-start life. The nitrogen cycle goes on and on. People who visit landfills are struck by their suggestion of death, but the landfill is also a place of resurrection: gulls and other wildlife live well off food scraps, and human scavengers, if they are allowed in, give new life to objects carelessly tossed—from furniture and clothing to building materials and VCRs.

In the winter of 2002, the *New York Times* ran a half-page story on an art exhibit in SoHo. The installation was called Cloaca: it consisted of a thirty-three-foot-long machine that replicated the human digestive system. A cloaca is a conduit that carries away sewage or surface water; it's also the common cavity into which the intestinal, genital, and urinary tracts open in vertebrates such as fish, reptiles, birds, and some primitive mammals. I read that twice daily an attendant fed Cloaca's mouth nutritious meals donated by fancy restaurants, restaurants apparently unafraid of being associated with shit. Twenty-seven hours later, Cloaca's back end excreted fecal matter onto a conveyor belt. I told Lucy about Cloaca, and she was eager to see it.

The New Museum of Contemporary Art was crowded when we arrived. Cloaca wasn't due to defecate for another twenty minutes. Lucy could barely contain her excitement. "Where will it come out?" she asked. She claimed a seat on the floor near the anal sphincter and stared at the glass vats, pumps, pipes, and tubing. Enzymes were injected from small tanks; liquid was siphoned off into a separate container. Cloaca wasn't all that different in function from the Owls Head treatment plant.

I was surprised to find the entire machine encased in Plexiglas. I asked the guard why. "When it first came it wasn't enclosed, but people complained," he said. "Oh boy, you shoulda smelled it in here." When I later asked the artist, Wim Delvoye, what had happened, he said the change was made at the demand of the museum because "the staff feared for its health. It's a pity, because the smell in fact is part of the piece."

When Cloaca finally defecated, about ten minutes late, Lucy was disappointed. "That's it?" she said. "It's not very big."

"No, it isn't," I said. "It looks like something you'd find in a litter box." I asked the guard what they did with it.

He said they threw it away.

Part Four

Piling On

Chapter Twelve

It's Coming on Christmas

It was nearly Christmas now, and every day the postman brought holiday catalogs. In 2001, American companies sent out seventeen billion of them—fifty-nine for every man, woman, and child in the United States—weighing a total of 7.2 billion pounds. Only six of the seventy-four catalogs surveyed by the advocacy group Environmental Defense used recycled paper in the body of their mailings. Switching to just 10 percent recycled content, the group said, would save enough wood to run a six-foot-high fence across the country seven times.

Dutifully, I sent letters and telephoned "opt out" services that removed my name from mailing lists. But what about the green groups? They (including Environmental Defense) kept pitching me credit cards. "Use our affinity card to make purchases," they said, "and we'll give a percentage to save herons or hemlocks." It made no sense to me: why did organizations that purported to understand the limits of our natural resources, never mind the problems inherent in waste disposal, make it psychologically and financially easier to buy more stuff? I expected green groups to critique consumerism, not promote it. (The cards themselves are made of PVC, which releases harmful synthetic chemicals—including dioxin, a carcinogen and hormone disruptor—into the environment during both production and disposal. According to Greenpeace, which pitches its own card made of Biopol, a "benign plastic made of cereal plants," consumers in the United States scrap fourteen million Visas and MasterCards a year.) Order something from those catalogs and it arrived swaddled in plastic air pillows, blister packs, or polystyrene peanuts and double-wrapped in cardboard. Onto the recycling pile went the paper, but the other material went straight to the dump.

Despite the many thousands of curbside recycling programs that accept paper, paper and other packaging waste still account for between 35 and 40 percent of the household waste in North American landfills. Americans don't care enough about recycling, it seems, and packagers have all kinds of incentives for wrapping things up in paper and plastic: to prevent theft (shoving a CD in a foot-long plastic bubble down your pants is a lot trickier than secreting just the disk), to facilitate self-service, to protect products from tampering, to provide a canvas for stickers that say "New!" or "Improved!"

Even the righteous European Union, which has a 53 percent overall recycling rate for paper, metal, plastic, and glass, is unable to keep pace with the growing tide of packaging materials. Waste generation is linked to economic growth, said a 2004 European Environment Agency report (no surprise there), and waste is increasing as more and more food, packaged for long-haul transport and longer shelf life, moves throughout the union. (In Europe, as in the United States, hyperwrapped convenience foods are becoming more common as potential household cooks go outside the home to earn a living.) The agency also cited the rising emphasis on health and safety for a surge in food packaging.

During a recent visit to Rome, where sanitation workers in stylish jumpsuits daily emptied the mini-Dumpster outside my friends' apartment building (at the laid-back Mediterranean hour of two-ish), I was struck by how far things had come. Twenty years ago, a Roman baker had handed me a bare-naked pastry: today, that pastry was placed on a cardboard tray, swaddled in more paper, and then hygienically bagged. My hosts' kitchen trash can was twice the size of my own, and they needed it.

On my first day in Italy, my friend dragged me, kicking and screaming, to a McDonald's for lunch. The largest fast-food chain in the world, McDonald's had in recent years tried to clean up its garbage act: it bought recycled paper, switched to lighter-weight packaging, and made big public donations to Environmental Defense. But the chain's waste, which was nearly as standardized as its food, the world o'er, was still staggering. In his documentary film *Super Size Me,* Morgan Spurlock ate only McDonald's fare for an entire month. During his McDiet he saved all his McTrash; he even mailed it home to himself while traveling. The result was thirteen large sacks—a six-foot-tall pyramid—of paper cups, lids, straws, ketchup packets, napkins, paper bags, burger wrappings, ice-cream cups, and french fry boxes. (Pressured by well-meaning consumers, Mickey D's gave up polystyrene clamshells for paper wrappings in the United States but continues to use them in other countries.) "Divide that by 90 (average number of meals) and then multiply by 46 million (number of people McD serves every day) and you get enough garbage to fill the Empire State Building," Spurlock wrote to me when I asked about his trash. "And that's just one fast-food company in one day." (It was also just the greasy detritus dumped on customers: Spurlock wasn't counting the bulk packaging left behind the counter.)

In a 2002 survey by *Packaging World* magazine, only 30 percent of the respondents, who made food, personal care, and pharmaceutical products for consumers, said that environmentally friendly packaging was "very important." When asked if consumers would pay a premium for green packaging, 61 percent said no. In the developed world, customers expect premium goods to come in premium packages. Cosmetic companies, which charge a lot of money for very small products, were the first to understand that cool packaging bespeaks cool contents.

I wasn't immune to the allure of the shiny and new, especially when it came to clothing, which I believed had some transformative power. It was the ceaseless marketing of, and the status seeking through, new possessions that I found anathema. The United States consumes far more stuff than any other developed country. According to the biologist Edward O. Wilson, if the rest of the world consumed at our levels—with existing levels of technology—we'd require the resources of four more planet Earths. (This extrapolation raises a question: before the last megafills are full, will we run out of stuff to put in them?) According to the United Nations "Agenda 21" report, "The major cause of the continued deterioration of the global environment is the unsustainable pattern of consumption and production, particularly in industrialized countries."

The flip side of consumption and production, of course, is wasting—consigning expired and unwanted goods to their fates in a landfill or incinerator. Why is there so much of it? Scholars offer various reasons. There is functional obsolescence (brought about by technical improvements); there is style obsolescence, also known as fashion; and the plain economic fact that it is often cheaper to buy something new than to repair something old. Wendell Berry wrote, in 1987, "Our economy is such that we cannot 'afford' to take care of things: labor is expensive, time is expensive, money is expensive, but materials—the stuff of creation—are so cheap that we cannot afford to take care of them." A lack of connection between those who make goods and those who use them contributes to the ease with which we turn our backs on our possessions. It is easier, for example, to throw out an ugly ceramic pitcher made in a Taiwanese factory than it is to throw out an ugly ceramic pitcher made by a well-meaning aunt or even an anonymous local craftsperson. Increasingly, handwork is not part of the equation.

The holidays are a perfect time of year to get across the source-reduction message. According to Inform, the environmental research firm, Americans produce an additional one million tons of trash per week between Thanksgiving and New Year's Day. (Waste watchers in college towns, however, note dips in garbage generation during this period and attribute it to the departure of students. Those cities' trash spikes come when students leave in late spring.) I found it hilarious that the city of Austin, which charges residents for each bag of garbage they set on the curb year round, offers a post-Christmas amnesty. "So you can throw out as much as you want and not be penalized for celebrating a Christian holiday with intense commercialism and the attendant cast-off crap," my Austin friend Spike Gillespie said to me. I knew that my city's holiday trash would find a place to settle, that it wasn't headed on a winding *Mobro*-style journey. There was plenty of room in the nation's supersized landfills. Even Fresh Kills had twenty years' worth of space when it closed. But in *Natural Capitalism,* author and entrepreneur Paul Hawken notes that for every 100 pounds of product that's made—product that hits the store shelves—at least 3,200 pounds of waste are generated. According to William McDonough and Michael Braungart, in *Cradle to Cradle,* "What most people see in their garbage cans is just the tip of a material iceberg: the product itself contains on average only 5 percent of the raw materials involved in the process of making and delivering it." In other words, we throw out stuff just to make the stuff we throw out.

It's been said that to make a dent in our garbage problems, source reduction has to acquire the rhetorical currency of recycling. One could argue that the currency of recycling isn't exactly robust, but at least it doesn't fly in the face of the holiday media messages one encounters at every turn. Inform offers loads of "greener" holiday tips, but they seem timid, dull, and rote. Give rechargeable batteries, a low-flow showerhead, a membership in an environmental organization. Yawn. Shop at thrift stores, send electronic instead of paper holiday greetings. Ho-hum. Clearly, source reduction has a PR problem. Compared to the sirens of high-end emporiums luring us to buy, Inform is a gray-haired spinster with an admonishing finger.

If the reality of an environmental conscience (as opposed to the idea of it) isn't chic, at least it is wholesome: it speaks to a connection with the natural world. We are all, pollutocrats and composters alike, children of the universe. And nature, as everyone knows, brooks no waste. That notion gets a lot of play in liberal recycling and design circles. "Consider the cherry tree," write McDonough and Braungart. "It makes thousands of blossoms just so that another tree might germinate, take root, and grow. After falling to the ground, the blossoms return to the soil and become nutrients for the surrounding environment. Every last particle contributes in some way to the health of a thriving ecosystem." It happens on every scale: large animals die, and their carcasses feed smaller animals, fungi, and microbes, which in turn make possible the feedstuff of larger organisms. The building blocks of life cycle endlessly, just like the nature shows that deliver this message without cease.

But do animals actually produce trash? Well, leaf-cutter ants carry dead workers from their underground nests, tipping them onto outdoor mortuary piles. They haul dried-out leaf fungi from garden chambers down trails to compostlike piles. (Both are exploited for nutrients by insectivores and herbivores.) Underground, kit foxes feed their young throughout the winter months. They pay the piper in the spring, when an entire season's worth of bones, fur, and feathers has to be excavated and placed on the equivalent of their curb (these scraps become food or shelter for other creatures). Prairie dogs, like humans, continually drop skin flakes and discard food in their burrows. Fleas, mites, lice, and other bottom-feeders constantly vacuum this stuff up: when the live-in maids become too populous, the original inhabitants clear out. Perhaps the animal most like humans, in its dealings with trash, is the pack rat. The desert rodents live in five-foot-high dens built of just about anything they can find (including branches, cactus pads, newspapers, cow pies, cans, and rags). When they periodically clean house or renovate, they create trash middens nearby. Varnished by coat after coat of pack rat urine, these heaps persist for ages in the dry desert. Just as William Rathje and his archaeology students analyzed landfills to suss the habits of Americans who lived a half-century earlier, scientists at the Desert Research Institute, in

Nevada, examine the radioactive fingerprints of crystallized pack rat urine and scrutinize seeds and pine needles to learn about climate and plant communities up to 25,000 years ago.

Could humans learn something from nature's constant cycling of resources? Sure, but there is a crucial difference between Homo sapiens and other species. In the natural world, no organism self-sacrifices for the good of the environment. Plants and animals steal whatever resources they can, heedless of fellow creatures and the future. Nature is both terrifically efficient and terrifically wasteful, all at the same time. Green thinkers like to focus on just one aspect of this equation. It's a feel-good idea. But nature isn't wise or far-sighted. Recycling, however, is wise precisely *because* it's far-sighted. Unfortunately, it isn't likely that we'll become truly efficient about resource recovery until we've exhausted all our raw materials (at which point the planet will be a fairly dismal place to live). Recycling is the name we've given to resource recovery before it's profitable. Only later, when recovery becomes a profitable necessity—because all the new material is gone—will we really be living like the animals.

The puffy winter coats we all wore as we made our shopping rounds—some jackets filled with feathers, others with flaked polyethylene terephthalate—only slightly disguised something else I had been thinking about. According to the Centers for Disease Control and Prevention, roughly two-thirds of adult Americans are overweight, with 22 percent qualifying as obese. For those who design and sell consumer goods, this is a whole new marketing niche to exploit. To accommodate bigger people, manufacturers make bigger clothes, couches, movie seats, and food portions. Doors are wider; so are church pews and caskets. Even regular-size people are buying bigger houses, television sets, and cars. Luxury recreational vehicles, with king-size beds and fireplaces, stretch to forty-five feet; houseboats—equipped with trash compactors—have reached 2,400 square feet. Architects design medicine cabinets as large as walk-in closets. Of course, all this bigger stuff requires more raw materials to make it, more fossil fuel to transport it, and more space in which to ultimately bury it. With every new product, more trees are cut, more metal is mined, more fuel is extracted and then burned.

Persuading Americans to consume less stuff, probably the single best thing we could do to save the planet (besides promoting energy conservation and zero population growth), isn't a big part of the environmental agenda. Instead, we are exhorted to buy green. Buy products—cleaning solutions, building materials, organic socks, paper goods—that are free of toxins in their manufacture, use, and disposal. Buy products that don't generate greenhouse gases, that can be refurbished and reused, that are minimally packaged. Green purchasing tells us to vote with our wallets, but it ignores a third choice: not buying at all. I resist the green buying message because I hate to think our strength is based in consumption, not in moral clarity.

The Buy Nothing message is, at any rate, politically taboo, and has been for some time. In 1930, the editor of *House & Garden* wrote, "The good citizen does not repair the old; he buys anew." To shop is American: to forgo consumption, unpatriotic. (These ideas were linked explicitly after September 11, when President George W. Bush urged the country to get out and buy, buy, buy.) Source reduction, the first in the three-R hierarchy, receives little attention, perhaps because it implicitly critiques the production side of the economy.

But consumers *can* make a difference. Frito-Lay, Heinz, and Kraft have added organic products to their lines solely in response to consumer demand. Pressured by custodians and other large-scale users, manufacturers of cleaning supplies have switched to less-toxic ingredients. But eating organic food and cleaning without dangerous chemicals are primarily perceived as health issues, with their environmental benefits only secondary. We are all used to reading nutrition labels, but what kind of nutcase checks the percentage of pre- and postconsumer recycled content on a package or whether its plastic container is recyclable in their community curbside program before committing to buy?

"You'd be surprised," Arthur Weissman, president of Green

Seal, a nonprofit environmental audit firm, told me. "People are very concerned about landfills spilling over." (Then again, he added, they also load their SUVs with recyclables to drive the ten miles to the transfer station.) According to Natural Business Communications, a consulting firm that tracks LOHAS consumers (that is, folks who practice "Lifestyles of Health and Sustainability"), sixty-three million American adults spent $226 billion on eco-conscious goods in 2000. I wanted to like those products, but sometimes I lost patience with their manufacturers' piety. They were still producing, marketing, and packaging a lot of overpriced, nonessential goods, then sending them around the country on trucks that burn fossil fuel. It was the treadmill logic of capitalism applied to a freshly identified green market segment.

In April of 2004, Wirthlin Worldwide, a consulting firm, reported that 82 percent of the consumers it surveyed said "corporate citizenship" had at least some influence on their buying decisions. But answering a marketer's survey is easy: spending the extra dough is hard. Seventh Generation, manufacturers of "earth friendly" household products, holds the barest sliver of that category's market share. Fair Trade coffee accounts for only 5 or 6 percent of premium coffee sales. Tom's of Maine claims about 2 percent of the enormous toothpaste market. (The craze for disposable cleaning products impregnated with harsh chemicals, meanwhile, seems to be spreading: the supermarket shelves are crammed with one-use toilet cleaners, dish-washing pads, dusters, floor sweepers, and mops.) Getting consumers to think systematically about their impact on nature, let alone landfills, isn't easy: the most immediate hurdle, for most shoppers, is the higher price of "green" goods. What spurs them over the hurdle is guilt about their planetary impact. "A brand can help us feel good," said Marc Gobé, author of *Citizen Brand*. "If you buy this yogurt, you don't have to make any other effort. You just buy it."

It is the same with scrupulous recycling. I could spend all day wrapping up my free AOL discs for a recycler in Green Bay, Wisconsin, organizing my polystyrene peanuts for another niche buyer, and baling my unwanted textiles to be shredded for felt. But these activities—virtuous as they might make me feel—are mostly a distraction from far more threatening issues. According to the Union of Concerned Scientists, which made exhaustive studies of consumers' environmental impacts, the things that make the biggest difference to planetary health are transportation, housing, and meat eating. It isn't worth it, they said, to get worked up over paper versus plastic at the grocery store. (For the record, the bags come out nearly equal in life cycle analyses.) The perfect, in the realm of recycling, is the enemy of the good.

"Don't worry or feel guilty about unimportant decisions," concluded the UCS. "Buy more of those things that help the environment." Does cruelty-free Aveda lipstick count, or Body Shop bath gel? No. But low-flow showerheads do, and Energy Star refrigerators, and hybrid cars, or converting your regular rig to biodiesel. So does buying anything that has already been recycled and that substitutes for something new, including aluminum cans, paper products, plastic wood, and toothbrush handles extruded from used Stonyfield Farm yogurt cups.

On my first post-Christmas pickup day, I ran downstairs to watch the garbage fairies at work. It was a new crew, and the guys told me that their routes, so heavy this week, had been halved. (Winter storms compounded the problem: because san men had been deployed with snowplows, our neighborhood had missed a trash pickup.) One of the workers, a weary-looking blond, sidled in between the parked cars and climbed over a dirt-crusted mound of snow. "Some people cut a little trail from the street to the curb for you," he said. "But then there's always someone who blocks it with a car." (Other residents formed the opposite of a trail: they piled sidewalk snow into an unhelpful ridge parallel to the curb.) He heaved a bloated black sack into the hopper and returned for a television. As he lifted it over his head, he slipped and fell on his back, hardly bobbling the TV. "You okay?" I asked. "Yeah," he said. "Happens all the time."

Earlier, I'd asked my regular guy Billy Murphy about seasonal changes in garbage. "In January, the weights go down," he said.

"There's less to throw out because there's no money left." I'd wanted to pump my fist in the air when I heard his simple analysis. It didn't take a lot of garbage theorists and environmental pointy-heads to make the connection between having a lot of money and discarding a lot of stuff. In general, more wealth equals more waste. (If Nantucket didn't have a median income of around $75,000, it might not need a digester to divert waste from its landfill. A major fraction of the island's trash is yard waste, from landscaping, and construction and demolition debris, from home renovation.)

I snooped around the 'hood, noting toy boxes, wrapping paper, envelopes from gift cards. I saw evidence of new clothes, kitchen equipment, tools, linens, furniture, office supplies. Spilling from trash cans and blowing down the slushy streets, the detritus fueled my spiraling negativity. Then suddenly, spurred by nothing tangible, I experienced a change of heart. The hundred million tons of refuse compacted under Fresh Kills' waving fields of fescue didn't have to speak of loss and greed, of excess and ignorance. Fresh Kills could stand also as a monument to generosity, to kindness and indulgence. We are creatures with aesthetic sensibilities: we surround ourselves with nice objects. The contents of the landfill say that we are rich, we have choices, we can afford to buy new sheets when we want to, not just when we need to. A century ago, refuse was an issue of poverty; now it is a sign of abundance, of economic vitality. A grubby doll's arm poking from a trash can wasn't sad, in this view. It was happy. The doll had been offered to a child out of love. The doll had *been* loved, and tossed away only when it was outgrown or broken.

The killjoy in me hissed, "Hey, why don't you fix up that doll and pass it along?" The mellower me shrugged and said, "Hey, let it go."

Until very recently, New York's Department of Sanitation sent dedicated trucks around the city to collect old Christmas trees from the curbs. Perhaps residents didn't understand, or care, exactly where their Scotch pines were headed, because many set them out wrapped in plastic, with skirts, tinsel, and decorations still dependent. The trees were bound for the chipper in Fresh Kills' compost yard, but Robert Lange, in the department's Bureau of Waste Prevention, Reuse and Recycling, had never been satisfied with the program, which cost $1.5 million. "The packer trucks would compress all these trees, and they came out in fifteen-by-twenty-by-ten-foot cubes. You couldn't pull the trees apart. A crane moved it to a chipper. The end product was not a useful horticultural product." This year, because of budget cuts, DSNY planned to collect trees along with regular waste and truck them out of state.

The first Saturday after New Year's, I joined a group of volunteers in front of Katina's, a local coffee shop, where a large red packer truck on loan from a private paper-carting company idled. The volunteers were mostly young, wearing beards, natural fibers, and "No War on Iraq" buttons. Our plan was to collect Christmas trees from the curb and bring them to Prospect Park, where they'd be chipped into earth-nourishing mulch. The city would save on tipping fees, and those far-off landfills would save some space.

The volunteers began dispersing around the neighborhood to pluck trees from the side streets and drag them to the avenues to meet the red truck. Who would stay with Adrian Istrefi, the driver of the big rig, and act as a liaison? "Me!" I called, excited to finally ride in a garbage truck. (DSNY let no one but employees ride their fleet.)

We headed north on Seventh Avenue, scanning the curb for Christmas trees. I was filled with a sense of do-goodism. We were rescuing organic material bound for the landfill! We were saving the city either $66 or $105 a ton on export costs, depending on whom you asked and when! We were going to capture the trees' nutrients and recycle them into a valuable horticultural product!

When we didn't spot any trees on the first block, I was sanguine. By the sixth block without trees, I was beginning to feel discouraged. Silly, even. But I tried to keep my spirits up. "It's okay," I told Istrefi. "It's early." I had a feeling he was just along for the overtime his company had offered. We were doing a good deed, but he had no illusions about its environmental impact: we were covering a minute fraction of the city.

We turned onto Union Street, where our eyes simultaneously locked on two discarded trees. I leaped from the cab with an apprentice's enthusiasm. "Whoa, take it easy," Istrefi said in his Rocky-like accent. He was half Italian, half Albanian, and he spoke four languages. One tree was small, and I easily tossed it into the hopper. We did the bigger one together. I wondered what the evergreens would be like "up top," in the houses where ceilings ran to fifteen feet. "Those people put out some enormous trees," John Sullivan, my san man, had told me. "They fill up your truck fast."

On the next block, Istrefi let me pull the cranks on the hopper—one for down and one for in. It made me feel tough but not particularly cool. Operating heavy machinery had never been a childhood fantasy of mine, as it had been for many san men. Still, I smiled while I cranked, and the folks on the street smiled, and the trees smelled really great, like a Christmas tree lot but more fragrant for the crushing. On President Street, a building superintendent pulled several firs from his basement for us. He seemed inordinately grateful, though I doubt he knew who we were or where his trees were headed. "Hello?" I called to a woman toting an infant in a backpack and dragging an eight-foot Scotch pine. "Can we have your tree?" "Sure," she said, looking confused. Everywhere, the Subaru station wagons were out in force, their roof racks laden with balsams and firs.

As the morning wore on, I got a minitaste of the san man's life. It wasn't easy getting in and out of the truck repeatedly. Istrefi bumped into a garbage can and spilled food waste all over the sidewalk. Dogs peed on the Christmas trees. I stepped in shit. Cars honked at us for blocking the road.

Istrefi stayed calm, never shifting above second gear. He compared the city workers with the privates as we trolled for trees. "There's competition and there's camaraderie between us," he said. "But city workers do six tons in eight hours. We *work*. They finish early and sit around. We go back out. At the dump, the city cuts in line. Because they work for the city, they think they own the city."

On a side street, we stood behind the truck and Istrefi smoked a Marlboro while we waited for a super to drag out some trees. I mentioned a story I'd recently read about a chain-smoking truck driver who accidentally ignited his load and burned for five hundred miles. "It's no joke," Istrefi said, unimpressed. "That happens all the time." Mostly it was cigarettes, he said, but sometimes batteries sparked themselves. "Once a guy behind me called my cell to say, 'I think you're on fire.'"

"What did you do?"

"I dumped my load and called the fire department," he said with a shrug.

On Third Street, we saw a white minivan filled to the gills with Christmas trees, topped by Christmas trees, and pulling Christmas trees behind in a travoislike arrangement. "I've got thirty-five," the driver told us. He was a Boy Scout leader, and his troop collected trees every year. We nodded approvingly, but I was jealous of his haul. And I felt a bit like an arriviste, coming late to the mulch party.

Istrefi and I headed east toward the park to empty the truck and got stuck behind Gloria Pabon, the only female san man at the Brooklyn 6 garage. She was dumping barrels and slinging bags into her truck but leaving behind the trees, on orders from the boss. After Pabon and her partner turned the corner, we got stuck behind a Parks Department pickup loaded with Christmas trees. "This is ridiculous," I thought, perhaps out loud. We were competing for trees with the Boy Scouts, the Department of Sanitation (workers had been told to leave the trees for volunteers today but to collect them on Monday), the Parks Department, and any nonallied individual dragging a conifer in an easterly direction. Our effort seemed puny and our enthusiasm naive. Our diesel truck—so big and shiny—was an affront to the grassroots spirit, to say nothing of the neighborhood's air quality.

At last we made it into the park. Istrefi hit a button, and his truck spilled its fragrant load onto the ground. "One hundred and nine, one hundred and ten," a volunteer counted while others dragged evergreens to a whining wood chipper. The area was crowded with happy Park Slopers, their children and dogs. Everyone was friendly, filled with a sense of environmental correctness and unlimited free cocoa. The Brooklyn Botanic Garden had set up an information table, staffed by a master composter. Though I was tempted to ask him about worms, I decided it would be best not to get started.

Istrefi went off to find more trees, and a U-Haul rented by the volunteers glided in and dumped forty-four more. An elderly gentleman with a homemade "Africart," a wagon bed atop bicycle wheels, exchanged his tree for a load of mulch. An extrovert in greenface held a sandwich sign: "Hey Buddy, Can you spare a pine?" "You want some mulch?" asked a young man standing by the chipped pile with a shovel. My compost bin could use the carbon, I thought, and I liked the idea of closing the loop. Then I got real: I didn't need to tote mulch home. I was making my own!

"No, thanks," I said, and turned to ask a Parks Department employee if this year was different from others. "It's the same," she calmly drawled. "People bring in their trees."

Later in the week, the *New York Times* reported that the sanitation workers' union wasn't pleased with the freelance tree-mulching projects. It didn't like other people doing the rank and file's job, and in what seemed a fit of pique, DSNY closed the mulching facility at Fresh Kills and refused to provide trucks to haul excess mulch from city parks to Staten Island. "It's stupid internal politics," Tom Outerbridge explained to me when I phoned. "The union is being immature and territorial." You mean the san men want to pick the trees up themselves, on an overtime shift, and then bury them? "Sort of," he said.

Then the other shoe dropped: Manhattan's DSNY-collected trees had been bound, along with most of the borough's solid waste, for the incinerator in Newark. But the incinerator, it turned out, wasn't permitted to accept "bulk vegetative waste."

"It's ludicrous that no one knew about the incinerator before,"

Robert Lange said when I gave him a call. So what did you do? I asked. "We sent out one round of trucks to get the pristine trees, the ones without tinsel and decorations, and bring them for composting to Fresh Kills, which they did open. Then we sent out another round of trucks to get the dirty trees." And where did they go? I asked, almost afraid of the answer. "To Wards Island, where Parks Department volunteers cleaned them."

There was a pause in our conversation. "And this is saving $1.5 million?" I asked, thinking of the latest budget cuts. Lange wouldn't answer on the record. "It sounds like a debacle," I said before hanging up.

"Yeah," he said. "It was a debacle. But it made people feel good."

Chapter Thirteen

The Dream of Zero Waste

It was trash night, and Willy—a grizzled man in his sixties—padded into the yard to peer inside our recycling can, dedicated now just to metal. If the bottle bill covered cat- and dog-food cans, Willy would have been in hog heaven, but tonight there was nothing here to interest him. I'd set out empty bottles of Samuel Smith's Taddy Porter and Samuel Smith's Organic Lager, but, like everyone else in the recycling world, Willy was particular about what he'd take. Discount Beverage, where he'd roll his overflowing cart at the end of the day, wouldn't accept what it didn't sell. I'd have to bring the beer bottles back to the food co-op if I wanted my two nickels back. I didn't, actually, but returning was the right thing to do—it supported the bottle bill and it kept weight off our city trucks.

Willy, who didn't offer his last name, made about ten dollars a day on the bottles and cans that Park Slope residents couldn't be bothered to return. That was two hundred containers plucked from the garbage. Willy thought there were a lot more bottles around since the glass suspension. "But they're harder to get," he said. "They're mixed in with the trash." Digging them out was a big hassle. Willy didn't like it, and neither did John Sullivan and Billy Murphy, who ended up with torn and slashed garbage bags.

It's axiomatic that whenever something is perceived to have value, it will be fought over. And in many communities with curbside recycling programs, glass *does* have value. It is mixed with asphalt for paving, or it is pummeled into shards and used as an aggregate in construction, or into even tinier shards for sandblasting or fiberglass, or it is color separated by an optical sorter and melted down into new bottles. Making new glass from old glass yields 50 percent energy savings over the scratch method, but fewer than a third of the glass bottles sold in the United States are recycled into anything, let alone refilled. Having long ago dismantled their refillable-glass-bottle operations, bottlers and distributors don't want to haul empties long distances back to their plants. Once upon a time, New York City's Department of Transportation paved roadways with glassphalt, but then it decided simply to recycle its old asphalt into new. With the elimination of glass from the recipe, the system was thrown into turmoil. Broken glass coated with peanut butter and spaghetti sauce began piling up in recycling centers.

"You could make a snowball out of it."

"I swear I saw a pile of it move."

"The rats," the gentleman next to me whispered in my ear.

We were sitting, more than a hundred of us, in a meeting room at Pace University, in lower Manhattan. The room had been set aside for a two-day recycling roundtable initiated by the Citywide Recycling Advisory Board, or CRAB, and the Center for Economic and Environmental Partnership. This was the Kyoto of recycling, a summit of industry types who wanted to make money off waste and avoid regulation, and environmentalists who wanted to keep waste out of landfills and incinerators. It had been about five months since the mayor had suspended plastic and glass recycling, and the meeting was one response to a city that seemed fresh out of solid-waste ideas.

For sixteen hours over two days, the participants discussed the minutiae of municipal recycling in a fiscal crisis. I heard about truck routes, labor unions, side- versus rear-loading trucks, single-stream versus dual-stream recycling, the effect of short-term contracts on capital improvements, pay as you throw, the virtues of dirty MRFs (materials recovery facilities that accepted recyclables bagged with garbage), the struggle to educate the public, and the scramble to claim unclaimed bottle deposits. Representatives from cities with landfill diversion rates in the 50 to 60 percent range talked about ways to increase tonnage, reduce contamination, whittle away costs, and boost revenue. Through almost all of it, the elderly representative from the Sierra Club slept.

Early on the second day, a student observing from the back row shouted out, "What about producer responsibility?" The room fell silent as industry reps looked around uncomfortably, contemplating a future in which any business that made and sold something agreed to take it back at the end of its useful life. The concept wasn't new: in 1993, the NRDC's Allen Hershkowitz had written in the *Atlantic Monthly* that "the nation's economy would be well served if municipal waste was reclassified as manufacturer's waste—and the waste itself became the financial obligation of the consumer-products companies." A chill settled thickly over the roundtable until a gentleman from the National Association for PET Container Resources bravely fielded the question: "We would not agree that sticking it to industry is fair." That got a laugh, and then we broke for lunch.

Robert Haley was the big cat at the roundtable, the guy all the media wanted to interview. Apple-cheeked and enthusiastic, with a cap of short dark curls, he ran San Francisco's recycling program, which had doubled its diversion rate—to 52 percent—while New York had slashed its program in half. San Francisco, Haley told the rapt audience during his PowerPoint presentation, was aiming for 75 percent diversion by 2010. In a few months, the city council would revise that optimistic goal, aiming for Zero Waste by 2020. From New York, San Francisco looked like a solid-waste utopia, with trucks that run on liquid natural gas, a massive composting operation, and happy Italian American workers making thirty dollars an hour, plus triple time on twelve holidays a year.

Before I left the roundtable, I dropped my plastic plate into the trash and chatted with Haley, who was momentarily free of admirers, about traveling through the garbage landscape. "San Francisco is full of kooks," he told me, grinning. "Now the city is thinking about using the tidal power of the bay. Come on out, and I'll show you our new MRF. It cost thirty-eight million and it's just about to come on line."

The MRF sprawled over 200,000 square feet on a pier in the Port of San Francisco. Approaching its front bays, I thought all the money must have gone inside, because the place looked like any other industrial warehouse in a grimy neighborhood of loading docks, cranes, and gravel mounds.

I had been touring San Francisco's garbage infrastructure for two days—prowling around the city's transfer station, poking into its curbside bins, and following its garbage trucks—and my sense of anticipation at seeing the MRF, coupled with jet lag, wasn't what it could have been. Still, I was having fun. Haley had passed me on to Bob Besso, who worked for Norcal, the private company with which the city contracted to pick up refuse. Dressed in blue jeans and sneakers, Besso had the lankiness of a marathon runner. He was in his fifties, and he'd worked in recycling for decades. His and Haley's easygoing attitude and their penchant for plain speaking were diametrically opposed to the formal inscrutability of New York's sanitation operatives. The best part of hanging around Besso was his competitive streak: both he and Haley, over at the Department of Environment, were walking poster children for Zero Waste. Who could throw out less? Who had more radically altered his lifestyle to leave a smaller human stain?

The Zero Waste concept was a growing global phenomenon. Much of Australia had committed to achieving the goal in 2010, and resolutions had been passed in New Zealand, Toronto, twelve Asia-Pacific nations, Ireland, Scotland, the Haut-Rhin department in the Alsace region of France, several California counties, and the town of Carrboro, North Carolina (also known as "the Paris of the Piedmont"). So far, no community had reached this nirvana, a condition perfected only by nature. For humans to achieve zero waste, went the rhetoric, would require not only maximizing recycling and composting, but also minimizing waste, reducing consumption, ending subsidies for waste, and ensuring that products were designed to be reused, repaired, or recycled back into nature or the marketplace. Zero Waste, said Peter Montague, director of the Environmental Research Foundation, had the potential to "motivate people to change their lifestyles, demand new products, and insist that corporations and governments behave in new ways."

I didn't take Zero Waste literally. I considered it a guiding principle, a rallying cry for green idealists. I understood its intensive recycling component, but what about goods that simply could not be recycled? I had in mind the Fuzzy Flower Maker hiding in Lucy's closet, a craft kit left at my house by a summer subletter. It was a hideous thing: you put glue on the petals of plastic flowers, then stuck them inside a plastic "swirl chamber" where a battery-operated spinner dusted them with colored powder. How was I going to get rid of this thing when the spinner broke or the fuzzy powder ran out? Surely the time it took to mold this toy from virgin resin, to assemble its unrepairable parts, and to ship it to a store was far longer than its working lifespan. The thing weighed just over two pounds (I was still weighing everything) and would seriously skew my garbage average for the week.

Over lunch in a Vietnamese restaurant, I learned that Zero Waste wasn't just rhetoric to Haley. "I don't have a trash can at work," he said. On his desk sat a grapefruit-sized ball of used staples—ferrous scrap that he couldn't bear to throw out. "If I'm going to be a leader in Zero Waste, I have to live the life," he said. I asked what effect this had on domestic harmony. "My partner is 99.9 percent with me," he said, nodding enthusiastically.

"What's the one-tenth-of-a-percent problem?"

"She draws the line at twist ties."

"Well, you know you could strip the paper from the wires

and—" I interrupted myself. Haley already knew how to recycle a twist tie. At home, he was diverting 95 percent of his waste from the landfill. The 5 percent he threw out was "manufactured goods"—recently some beyond-repair leather shoes. Worn-out sneakers, of course, were mailed to Nike, which shreds rubber and foam into flooring for gyms. The company accepts non-Nike footwear, too, and is also trying to tan leather without questionable toxins and developing shoes made of a new rubber compound that doubles as a biological nutrient—something that can be harmlessly returned to nature. This would be quite an improvement, since, according to William McDonough, conventional rubber soles are stabilized with lead that degrades into the atmosphere and soil as the shoe is worn. Rain sluices this lead dust into sewers, and thence into sludge bound for agricultural fields. According to the National Park Service, which has more than a passing interest in man-made stuff that lies around on the ground, leather shoes abandoned in the backcountry last up to fifty years (if they aren't eaten, one presumes), and rubber boot soles go another thirty. Aluminum cans, agency workers predicted, would last eighty to one hundred years, and cigarette butts and wool socks between one and five.

McDonough's 206-page book, *Cradle to Cradle,* was printed on "paper" made of plastic resins and inorganic fillers. The pages are smooth and waterproof, and the whole thing is theoretically recyclable into other "paper" products. The book weighs one pound, four ounces. A book of comparable length printed on paper made from trees weighs an entire pound less. "What do you think of that?" I asked Haley. He nearly spit out his mouthful of curried vegetables. "McDonough's book will be landfilled! I'd rather cut down a tree!"

To Haley and Bob Besso, landfilling was the ultimate evidence of failure. Avoiding the hole in the ground—which in San Francisco's case was owned by Waste Management, Norcal's archenemy—had become a game to them, albeit a game with serious consequences. Haley didn't use his paper napkin at the restaurant, and he scraped the last bit of curry from his plate. But we all knew there was waste behind his meal—in the kitchen, on the farm, in the factory that made the box in which his bok choy had been carted to San Francisco.

I wondered if Zero Waste really meant anything, considering the limits of our recycling capability and our reluctance to alter our lifestyles. It was as dreamy an idea as cars that run on water. And just as appealing to industry, too. "Zero Waste is a sexy way to talk about garbage," Haley said. "It gets people excited." I considered that for a moment. Could we solve our garbage problems by making garbage sexy?

Seeing how little I could throw out was fun for me, if not exactly sexy. I'd gotten caught up in the game, back home with my kitchen scale and Lucy's blue toboggan. I recorded my weights in a little book, I crunched my numbers, and I measured my success by how many days it took to fill a plastic grocery sack. Being handed a purple Fuzzy Flower Maker, in this context, was like picking up a Go to Jail card in Monopoly: it set me back several turns.

Faced with a child who coveted such a toy, I asked myself, "What would Haley do?" The answer was easy: he wouldn't have considered buying it. In the Zero Waste world, Fuzzy Flower Makers either wouldn't exist or they'd be made of "nutrients" that could be reused in some other form. (If Haley had children, and he didn't, he'd probably make them play with the biological nutrient called flowers, which come with their own colored powder—pollen.)

In the months to come, I'd find people who neither lived nor worked in the Bay Area who were having fun (if not sexy fun) with garbage reduction. Shaun Stenshol, president of Maui Recycling Service, had toyed with the idea of decreeing a Plastic Free Month, but ultimately deemed such a test too easy. Instead, he issued a Zero Waste Challenge. Over the course of four weeks, Maui residents and biodiesel users Bob and Camille Armantrout produced eighty-six pounds of waste, of which all but four (mostly dairy containers and Styrofoam from a new scanner) was recyclable. Alarmed to note that 35 percent of their weight was beer bottles, which they recycled, the Armantrouts vowed to improve. Bob ordered beer-making equipment to help reduce the amount of glass they generated, and Camille promised to start making her own yogurt. Despite these efforts, the Armantrouts didn't win the Challenge. The winner of the contest, as so often happens, was its inventor. All on his own, Stenshol had produced an even one hundred pounds of waste, of which he recycled ninety-nine.

San Francisco had one thing going for it that New York probably never would: a pay-as-you-throw system. It involved three separate curbside bins. The black one was for refuse; the blue was for recyclables (metal, plastic, glass, and paper); and the green was for "organics," which included yard waste, kitchen scraps, and food-stained cardboard. Weekly pickup cost $16.49 a month for a thirty-two-gallon container of refuse (the price dropped 20 percent if you used a twenty-gallon container), but recycling and composting were free. Eventually, the city would raise its garbage rates and start charging for recycling, but for now dumping milk bottles into the blue bin was free.

One of the chief objections to pay as you throw in New York City was the burden it placed on low-income residents. Was it fair for them to spend a larger proportion of their income on garbage disposal? "We have a reduced-rate program, but very few people have applied for it," Besso said. He didn't care: he was adamantly opposed to the program. "I think everyone should pay for their trash. What are low-income people throwing out? They're poor!"

It sounded harsh, but it is obvious to anyone who has contemplated the contents of a packer truck that people with a lot of money discard more things than people without a lot of money. Rich people remodel and upgrade; they buy new clothes and landscape like crazy. But poor people, I realized, are often guilty (by necessity) of buying cheap goods, stuff that falls apart and can't be fixed. I'd done it plenty of times. Instead of buying a more expensive shower curtain that could stand up to repeated washings, I'd buy a cheap plastic one and replace it in two years.

Pay as you throw had been proposed in New York City, but the

Department of Sanitation worried about residents dumping their household waste in the streets. More than six thousand communities across the nation—including big cities like Minneapolis, population 373,188, and small towns like Dog Patch, Arkansas, population 608—used some form of pay as you throw. And yes, they had problems with roadside dumping and with people sneaking garbage into cans other than their own. Besso had lobbied to remove litter baskets from San Francisco's street corners for just this reason, but the Department of Public Works won that battle. It galled Besso that the litter baskets were topped with separate bins for recyclables, which made it easy for street people to poach bottles for the five-cent return. "They're stealing from us," he said.

We were driving around town now, following garbage trucks and poking into garbage bins. In some neighborhoods, the bins were padlocked. "The green ones contain edible food," Besso explained. "We don't want people putting recyclables into the refuse, and we don't want people poaching the deposit containers from the recycling."

Behind a restaurant in the Embarcadero, Besso proudly pointed to a busboy tipping food scraps into green bins. Before the city had introduced organics into its recycling program, its diversion rate was close to New York's. Collecting the green stuff had raised San Francisco's rate by 15 percent, and a lot of that came from restaurants. The food scraps were trucked to Vacaville, California, a couple hours away, and composted at a landfill run by Norcal. "A lot of these restaurants buy produce that's grown in their own compost," Besso said. "So we're closing the loop."

We headed for Chinatown, which had the worst diversion rate in the city. "Their consumptive and disposal habits are unique," said Besso. "We lose ten to twenty percent of metal from residential bins in this neighborhood to poaching. The Chinese wrap their fresh food in paper, not plastic. We just don't get a lot of stuff out of them. The route is difficult and tight, and the people just don't care. They pay to throw it all out."

Norcal had adapted to some of the city's many peculiarities. On the narrowest streets, workers hunted down refuse in skinny trucks. In one neighborhood, they used rear-loading dual-bin trucks; in another, they used side-loaders. In the steep-sided neighborhoods near Coit Tower, they abandoned their trucks and went on foot. With burlap sacks, they hiked up and down hillsides fragrant with camellias and magnolias, collecting foie gras wrappings and fizzy-water bottles from maisonettes covered in vines. Each semiautomated truck had one worker, who picked up an average of eight tons of trash and recyclables during an eight-hour shift. In New York, by contrast, it took two men to collect an average of 9.7 tons per shift. (Toronto took the efficiency prize, with workers collecting, on average, twenty-five tons of trash and recyclables during their ten-hour shift.)

Curbside recycling isn't cost effective by the usual calculations. But like police protection and libraries, it is a service that the public—especially in places like San Francisco, Madison, Wisconsin, and New York City—has come to expect. No one requires garbage collection to pay for itself; why should recycling? John Tierney, who became the nation's recycling antichrist with his 1996 *New York Times Magazine* article "Recycling Is Garbage," argued that recycling cost more in labor than it saved and that there were no shortages of natural resources to make new paper, cans, and bottles. So why bother? The article generated more letters, two-thirds of them outraged, than any other in the magazine's history and spurred lengthy point-by-point rebuttals from the NRDC and Environmental Defense. Eight years on, Tierney has not recanted. "I made one mistake in that article," he told me. "I overestimated the value of real estate taken up by a recycling can in the kitchen." Still, he allowed that automation at MRFs might make some recycling programs economical. Until then, it wasn't worth his time, he said, to walk a soda can down the office hall to a recycling bin. "Materials keep getting cheaper, and labor is more expensive."

Besso poked his head down an alley to look for recycling carts but turned around when he heard the clang of cymbals. A group of Chinese men was marching down the middle of the street. Dressed in saffron robes with burlap hoods, they carried bamboo poles and swung incense. A large photograph, a black-and-white portrait, rode in the backseat of a convertible. "I've never seen that before," Besso said, staring at the burlap hoods. "Those look like our sacks."

The funeral procession passed, and we moved on to Pacific Heights, where we watched a sanitation worker open a gate underneath an elegant five-unit building. (For a price, workers would even come inside your home to collect garbage from your kitchen.) In an alley adjacent to the garage, both a ninety-six-gallon recycling bin and a ninety-six-gallon garbage bin were full, but the green bin, for organics, was empty. As I was turning back toward the truck, a young man in a cashmere jacket came whistling up the street. This was his house, his trash. "Hello," I said to him. "I was just wondering where your compost is."

"Compost?" he repeated.

"Yes, the green bin."

"I didn't even know we had a green bin," he said. "We have no incentive to compost. The landlord pays the trash bill. Besides, there's no room in my kitchen for another bag."

It was refreshing, in a way, to see that San Francisco's trash utopia was a little bit of a fairy tale. The city spent two dollars per resident on garbage education. New York, which spent about a dollar per person, was perennially criticized for shortchanging recycling education efforts. But San Francisco's Chinatown wasn't pulling its weight, and neither were the city's high-rises, where it was administratively challenging to enforce pay as you throw, physically challenging to manhandle tons of bulky recyclables, and entomologically challenging to store quantities of food waste. And in a crowded city where people were all too used to overpaying for their housing, it was easy to see how some would pay a little bit more—call it a Tierney tax—to free up an extra square foot of kitchen space.

Haley thought San Francisco would reach its goal of 75 percent diversion in 2010 by continuing to roll out its three-bin program in every neighborhood of the city. "We'd expand the recycling, go after packaging, mostly—food service and take-out containers," he said. "We'd also have to make recycling mandatory. If you don't recycle, we won't pick it up." Ultimately, though, achieving Zero Waste would require legislation that mandated high percentages of recycled content in bottles and other materials, as well as the dreaded (by industry) extended producer responsibility, in which manufacturers not only design products made of recyclable materials but also take back and reuse their packaging. In Germany, which is at the forefront of this movement, manufacturers are required to pick up discarded packaging at the point of sale—the supermarket or the department store, for example. The stores don't like it, but between 1991 and 1997, packaging declined by 13 percent, per capita, compared to a 14 percent increase in the United States.

That San Francisco managed to divert 52 percent of its waste today, I realized, was largely because it collected organics and because commercial establishments were included in the city's figures. (New York businesses were required to recycle, although the law was not enforced. Private carters handled commercial waste, and the DSNY didn't include their efforts in their figures.) Like New York's Department of Sanitation, Norcal collected bulk items from city streets and had drop-off sites for electronics and other items difficult to recycle. But unlike New York, San Francisco didn't landfill or incinerate those discards. A mattress, for example, was separated into its component parts: cotton and horsehair and springs and wood, all of which got reused.

"The labor costs as much as tipping it in the landfill," Haley said. "But that doesn't matter. We're making a point. We're taking a long-term broad view and trying to get to Zero Waste." We were back at Haley's office now, sitting around a table with a centerpiece of three tiny plastic trash bins: black, green, and blue. I couldn't help thinking that they'd ultimately end up in the dump. On my right was Virali Gokaldis, who was writing a report on recycling for the Natural Resources Defense Council. A policy analyst, she had been taking painstaking notes on how much the city paid to collect waste and process it. Over and again, she asked Haley for the breakdown of costs and rates and savings. Haley provided no numbers, but he repeatedly spread his hands four feet apart and declared, "I've got this many folders on the shelves in my office." It was unclear whether Gokaldis was being invited to pore over these data, but the momentum was stomped by Haley's boss, Jared Blumenfeld, who now said, "Shit. If we tried to figure all the costs of recycling a mattress we'd never get anything done. We say, Do we want to landfill this thing or not?"

As one, Besso, Haley, and Blumenfeld shouted, "No!"

Vast and tumultuous with the roar of front-end loaders and the crash of breaking glass, San Francisco's $38 million MRF loomed before us. Avoiding packer trucks on the tipping floor, I sidled along a wall. Like my transfer station on Court Street, the MRF wasn't designed to accommodate visitors. There were no waiting rooms or receptionists, no viewing galleries or carpeted hallways. The place was all business.

The MRF appeared to be one single contraption, a sprawling machine composed of hoppers, balers, and bins connected by belts, chutes, and ducts. The place could handle up to fourteen hundred tons of material a day, though it currently processed less than half of that. From where I stood, everything appeared to be in motion: eighty-seven different conveyor belts were lifting commodities up, dropping them down through chutes, and sweeping them through finger screens, disc screens, and human hands. A blast of air floated paper toward six "lines," where workers sorted and fed it into vacuum ducts leading to balers. Magnets pulled out steel; eddy currents spun out aluminum. The air smelled pleasantly of beer.

I climbed a metal staircase to an elevated meeting room that overlooked the plant. The floor in here was carpeted and clean, which immediately set this MRF miles above any other garbage facility I'd visited. I gazed through plate glass windows and made out, against a dark wall, a line of workers plucking plastic from a conveyor belt at waist height. They toiled inside wire cages, protected from flying debris. "Could I walk down and see what they're doing?" I asked a plant manager. "Absolutely not," came the answer. I wanted to see what the workers—all of them African

American—took from the line and what they ignored. Denied access, I wondered if there was a lot of reject material or if the job was dangerous. I knew that worker turnover on the line was high. The job seemed only marginally less miserable than the metal-picking job, handled by elderly women working a conveyor belt, at Hugo Neu. This was one of the dark undersides of the green revolution, I thought: everyone wanted to recycle, but nobody wanted to do the recycling.

When the MRF had first opened, a few months earlier, workers separated plastics numbered one through seven. Now they focused on just two categories: PET and everything else. I asked the manager if he had to reconfigure a lot of equipment when the object of desire switched from HDPE, for example, to LDPE. "Nope," he said. "We just tell the people on the line to do it." Thirty-eight million dollars, I thought, and still it comes down to that.

PET was the most valuable plastic that rolled through the MRF. It was sold at a fairly steady rate to a broker, who in turn sold it to clothing manufacturers. A company called Epic, in the East Bay, shredded San Francisco's plastics numbered 2, 4, 5, 6, and 7 into pellets they melted and extruded into lumber. Sometimes, if the market was strong, number two plastic was sold for sleeping-bag insulation. Number three plastic, aka PVC, was the demon seed of recycling, a contaminant to almost any batch of melted plastic it touched. Norcal collected it at curbside and drove it to the MRF. Mixed in with other grades, it soon became Epic's problem.

Besso made it sound as if Bay Area plastics were, for the most part, recycled locally. But according to California's Integrated Waste Management Board, the majority of PET plastics and about half of all the HDPE collected within the state in 2001 had been exported. San Francisco's diversion rate may have topped 50 percent in 2003, but California overall was actually fifth in the nation that year, recycling at a rate of 40.2 percent. In first place came Maine (at 49 percent), followed by Oregon (48.8 percent), then Minnesota (45.6 percent), and Iowa (41.7 percent). Apparently,

California was little different from the nation as a whole: it contained a myriad of recycling microclimates that changed from moment to moment. What was eagerly collected in one community was pure garbage just over the county line, and the rules could change overnight. The rainmaker was the buyer, helped along by municipal finances and political will. But mostly, successful recycling depended on markets.

Without getting down among the San Francisco MRF's conveyors or poring over a schematic, I found it hard to follow a load from the tipping floor through the sorting lines to its proximate destination: bins and bales. I had to be content with pressing my nose to the back window of the conference room and watching as lines of cans and paper, now combed into single strands from the tangle up front, rolled into a hopper. Iron walls closed in on each load, cubing it. A whiplike arm wrapped the bale in wire, and a pushing blade shoved out, like meat from a grinder, a nearly continuous stream of four-by-four-by-six-foot rectangles. Forklifts buzzed around the back half of the MRF, arranging 1,550-pound bales of paper and 1,000-pound blocks of shiny aluminum cans—red, silver, and blue—into walls that rose twenty feet. Pigeons swooped from the rafters. Rats crept from the shadows and, like tourists among ancient ruins, picked their way through narrow passageways.

When New York sold its paper and its metal, the city profited. When San Francisco sold its paper and its metal, Norcal did. "Selling the recyclables is just cream for us," Besso explained. His company was paid through garbage collection fees and was guaranteed, by the city, a certain rate per ton; if it could supplement that income by selling any part of that waste, then more power to it. "Don't forget we're paying to tip at the landfill in Altamont, so we want to tip as little as possible," Besso added.

Every six to eight hours, a tractor-trailer left the Port of San Francisco for the landfill, laden with broken glass, rubber bands, plastic milk jug tops, and twist ties, among other items. Taken together, these scraps and orphans were the moral equivalent of my purple Fuzzy Flower Maker. How could we achieve Zero Waste in a world where such things fell through the cracks, where no category (or, to be precise, no market) existed to catch them?

Glass was the most musical of all recyclables. It tinkled steadily down through cracks and onto conveyors with a sound not unlike wind chimes. Unbroken bottles sorted by color provided a steady bottom counterpoint as they rolled, *clunk-clunk,* into bins. Norcal sold the broken stuff to a middleman, who sold cullet for road base, fiberglass, and Gallo wine bottles. Any bottles that completed the journey from the packer truck, through the manipulations of the front-end loader, and out the other end of the MRF in one piece were rewarded with renewed life as a beverage bottle. "Even using cullet," said Besso, who'd gotten his start in the business washing bottles, "it's cheaper to make a bottle with old glass than from silica. You use less energy because the glass doesn't have to be as hot."

In 2001, nineteen states had dropped glass from one or more of their curbside programs, but other cities forged on. In Portland, Oregon, residents protected their glass bottles and jars from breakage by placing them within paper bags inside their recycling bins; san men placed these bags in a separate compartment of their recycling truck. Seattleites also kept glass apart from their metal, paper, and plastic: it was pulled out at the MRF, then optically sorted and sold to a local manufacturer of glass containers. In cities with single-stream recycling, like Phoenix, paper buffered the glass. San Francisco was still experimenting with different compaction rates, looking to keep bottles and jars in one piece. California and Oregon had strong incentives to recycle: the states had established minimum-content laws for glass products, including beverage containers and fiberglass insulation. In California, said Besso, this law drove the demand for glass cullet, whether mixed or single color.

I wondered if Besso was jealous of Seattle. He had recently sent me an article headlined "Emerald City Isn't as Green as It Used to Be" and warned me that the only reason Seattle's rate of diversion from the landfill looked so rosy—rosier than his city's—was because residents sent so much yard waste to a compost facility. I called Rebecca Warren, who lived in Seattle, to see what it was like living in a recycling mecca, a place on the verge of collecting commercial food scraps for a digester and banning all recyclables from household trash and all paper from commercial waste.

"Every Wednesday morning," Warren told me, happy to discuss her detritus, "a truck comes for my trash. Every other Wednesday, another truck comes for two different recycling containers. One holds metal, plastic, paper, and gable tops [those peak-roofed cardboard cartons used for milk and juice]; the other is a bin for all colors of glass. On the nonrecycling Wednesday, another truck comes around for yard waste. But I don't use that truck." It's not that Warren didn't have yard waste—she had plenty. She just didn't want to give it up.

Instead, she shoved her branches and garden stalks into a four-foot-long composting cylinder. Her leaves, and her neighbors' leaves, went into a pile for the cultivation of leaf mold. "Leaf mold is killer," she said, meaning it made excellent mulch for her extensive garden. Debris from that operation ended up in either another random compost pile or, if it was thorny, thick-leaved, and "horrible," in her truck, which she'd eventually tip at the town compost site. There, it would be transformed with the rest of Seattle's yard waste into Cedar Grove compost and sold throughout the region.

I thought Warren was finished describing her domestic scrap operations, but there was more. "I put some of my kitchen food waste into a trunk-sized worm bin, which produces a high-nutrient, fine humus," she continued, "and some goes to a 'green cone' composter, which looks like a flat-topped volcano with a below-ground leachate pit." The cone produced a lower-nutrient compost than did the worms, though with far less effort. Because neither method seemed to be consuming eggshells, Warren recently began collecting them as well, in a jar on her kitchen counter—a way station until she determined the shells' highest and best use. Twice a year Warren organized a clothes swap; once she brought an old mercury thermometer to a household drop-off site; she hadn't produced a drop of electronic waste in her life; and she placed her spent alkaline batteries in the trash because the city's recycling coordinator had told her it was okay.

Warren seemed slightly fanatic to me, especially with those eggshells, but I knew her heart was in the right place, because she regularly gave the sanest sort of environmental advice, under the pseudonym Umbra, to online readers of her *Grist Magazine* column. Though she lived in a city that enabled her recycling compulsion, Umbra constantly urged others to lighten up, to quit worrying about the tiny things. "Readers ask me what to do with a prescription pill bottle if their curbside program won't take it," Warren said to me. "I really think people have too many choices. Even the waste stream is confusing." Where the pill bottle went, she said, wasn't important. "They should worry about the big things, like transportation and housing. What are they driving? How big is their home heating bill?"

The following day, Besso collected Gokaldis and me, accompanied by two NRDC interns, and drove us north through the East Bay toward Sacramento. Within an hour we were in farm country. Freshly tilled fields alternated with freshly sprouted condominium clusters. Just south of Vacaville we turned east, where the sky was big and periodically shadowed by low-flying jets homing in on Travis Air Force Base.

Yesterday had been devoted to the contents of San Francisco's blue bins. Today we were on the trail of the green bins, the organics. The J&B Sanitary Landfill sprawled for hundreds of acres, but the part we were most interested in was the small portion called Jepson Prairie Organics. Jepson was a sister company to J&B, and they were both owned by Norcal, which was beginning to look as megalomaniacal, to me, as Waste Management. San Francisco sent fifty thousand tons of food scraps and yard waste to Jepson Prairie each year. "That's wet tons," said Greg Pryor, the landfill manager. He was bearlike and affable, standing six-foot-five, with wheat-colored hair, tobacco-stained teeth, and a beard that covered only the margins of his cheeks. Every now and then he put a chaw of tobacco in his mouth and spat into a polystyrene cup, a plastic for which there was currently no local recycling market.

Jepson received three hundred tons of food waste a day, some of which came from San Francisco residents, but most of which came from about eleven hundred institutions looking to lower their garbage bills. Accompanied by Pryor and Besso, I walked up a truck ramp to get a bird's-eye view of the same stuff I was tossing into my own compost bin. The food was dumped into a grinder, which spewed it into a conical pile on the ground. I saw a smattering of parsley tops, some orange peels, a Christmas tree, apples beyond their prime, cardboard, coffee grounds, and, after a slight zephyr came up, the BART card of one of the NRDC interns. "Oh, no!" she cried. "The screener will get that," Pryor said, unfazed by the possibility of contamination.

There was a chance that the eggshells and bread crusts arrayed before me were the remnants of foodie shrines like Zuni Café and Greens, places I longed to eat at. But the organic matter could just as easily have come from UC Davis's veterinary barns or even the state penitentiary at Vacaville, which made a daily drop. "You're lucky," Pryor said now. "The prison truck is just coming in."

We heard the *beep-beep-beep* first, then watched the dump truck back up to the bottom of the pile. "It doesn't dump from the ramp into the grinder?" I asked Pryor.

"You'll see," he said. The gate of the truck swung open, and its contents sloshed to the ground. The food, which was wet and predominantly orange, had been precut for knifeless prisoners into tiny bits.

"It looks like vomit," an intern observed.

"Yep," Pryor answered, spitting into his cup.

The pile smelled rich but not particularly garbagey at this moment. Still, insects have a keener sense of smell than humans, and that's why pheromone traps, which drowned flies in a ring of water, dangled under the truck ramp. Pryor also released parasitic wasps, which lay their eggs inside live flies. When the wasp larvae are born, they chew their way out of their host, a surefire way to end the fly's reproductive career.

A front-end loader arrived now to mix scoops of food with yard waste, then dump them into the hopper of a bagging machine. With a forty-eight-horsepower push ram, the machine stuffed would-be compost into a two-hundred-foot-long, twelve-foot-diameter bag made of black PVC plastic. "It's like stuffing a sausage," Pryor said. As the organic material inside the tube decomposed, the temperature climbed.

"We need to get 131 degrees for three readings in a row," Pryor said, in order to nurture thermophilic organisms. "From 90 to 120 degrees is the mesophilic range. Bugs like that. The pathogens are killed at 131. Let it get too hot and you kill the good ones." He regulated the temperature with a blower, which forced air through four-inch perforated pipes that ran the length of the tubes.

I strolled the muddy path between bags—there were eighteen of them cooking at once—and kicked at an escaped brussels sprout. I leaned against a bag. It was packed hard and it radiated heat like a furnace. A small ventilation hole gave off enough steam to warm leftovers. I felt a nearly irresistible urge to climb onto a bag and skip down the long fat row, like Dick Van Dyke on the rooftops of Mary Poppins's London. Already stray cats sprawled up there, basking in the by-product of biotic decomposition, oblivious to the overhead jets and the periodic blast of a shotgun, intended to scare birds from the landfill.

After two months in plastic, the compost was released into long windrows that would cure for another thirty days. An auger wormed its way down the row every three days, turning and aerating. At this stage, the compost looked rough—more like mulch and uncut tobacco than a fine loamy soil. It smelled earthy, of fungus and rot and mold, which took the shape of orange blobs and lacy white frills. It was the basis for new life, this evidence of death.

Unlike my front-yard composting bin, or Christina Datz-Romero's community composting project under the Williamsburg Bridge, Jepson Prairie was a full-on industrial operation. It looked simple enough, but it involved fleets of heavy equipment running on diesel and tearing up the roadways. I shuddered to think of the greenhouse gases and volatile organic chemicals, which contribute to the formation of smog, pouring from the vents in the tubes, but what, really, was the alternative?

Anaerobic digesters! Tom Outerbridge would have suggested. In fact, Norcal would soon be contemplating this tactic, and probably in a spot much closer to the city. Every day, new restaurants and homes were signing on to San Francisco's organics program—Jepson Prairie's output grew by 70 percent in 2003. But the prairie could accommodate only so many windrows. And doubling capacity, the company's goal, would put another twenty-three trucks on the road and pump an additional 1,020 pounds of VOCs into the air daily (938 more pounds than the county's strict Air Quality Management District allowed). Something would have to give.

For now, though, Jepson Prairie continued to cure and screen its compost, then sell it to local wineries, organic farmers, and landscapers at five to ten dollars a yard. "That's a good price, and it's good quality," Pryor said. "There's lots of nitrogen in it because of the food. It's black gold."

Before we left the area, we stopped in at Eatwell Farm, where Nigel Walker annually spread about five hundred yards of black gold over his seventy bucolic acres of lavender, fruit trees, and vegetables, some of which he sold to the Greens restaurant in San Francisco. We admired Walker's organic produce, our stomachs growling, and then squeezed into Pryor's truck. We were so hungry we started nibbling at the lavender honey and rosemary salt that Walker had pressed upon us. Then we started arguing about where to eat lunch. There weren't a lot of restaurants in nearby Dixon, and dietary restrictions further constrained our choices. Pryor had recently dropped from 330 pounds down to 308 after three weeks on the meat-centric Atkins diet, and he didn't want to lose his edge. Gokaldis, on the other hand, was not only vegetarian but vegan. We finally settled on Mexican, but only if the rice wasn't cooked in chicken stock and the beans contained no lard. It occurred to me that Nigel Walker's produce, while certified organic, had been grown with compost that contained meat, with its attendant hormones and antibiotics, as well as yard trimmings that likely contained fertilizers and herbicides.

We were driving past neat rows of almond and walnut trees now, and the conversation segued to the agricultural value of night soil. Jepson Prairie used to mix biosolids from the local wastewater treatment plant into its product, Pryor said, but once farmers started spreading the stuff on food crops, that practice had to stop. We wondered out loud whether Nigel Walker, an extremely earthy guy, would want biosolids on his fields. Pryor said, in the tone of someone offering inside information, "Nigel isn't really into using shit."

I pondered that for a moment. "I don't think he's exactly *anti*-shit."

"Yeah," Pryor said thoughtfully. "You'd think he'd be pro-shit."

"I think he's pro *his* shit," I said. We nodded in simpatico, then parked the car for lunch.

I didn't want to leave the Bay Area without visiting Urban Ore, the reuse and recycling center in Berkeley. The emporium, located in a low-slung industrial neighborhood near I-80, bought about 85 percent of its goods from individuals and wrote trade credits for the rest. Two-thirds of sales were building materials, which were stacked outside a corrugated warehouse in an asphalt yard. I saw hot tubs and sliders, wrought iron gates and shower stalls, medicine cabinets, lumber, brick, and pipe. It was like Home Depot, but nothing was shrink-wrapped or new. Inside, Urban Ore looked and smelled more like the Salvation Army, with racks of clothing, shelves of housewares, kitchen chairs, books, toys, and the usual complement of bric-a-brac. I saw stalls of bikes and toilets, tables laden with doorknobs, hinges, and LPs. I saw louvered shutters and appliances, a five-foot-wide Frigidaire in beautiful condition, a midcentury monster labeled "big ol' freezer." Urban Ore had been called the "Best Urban Junkyard" by a local newspaper, and its single editorial comment said it all: "Fuck Ikea."

I strolled around and watched contractors picking through faucets and moms flipping through clothes racks as Johnny Cash spun on a turntable. A sign read: "Wasting is obsolete." If something didn't sell it was broken down and recycled: bike wheels into metal and rubber, beach chairs into aluminum and plastic webbing.

What about the clothes that won't sell? I asked Dan Knapp, the center's founder. "We try to get textiles to handlers who wash and sort them," he said. "They either ship them to developing countries or sell them to boutiques or they shred them for fiber. They're turned into wiping cloths or cloths for cleaning up chemical spills." Knapp had introduced the concept of "total recycling" in a 1989 white paper, listing twelve master categories of waste. There were reusable goods (including intact or repairable home or industrial appliances, household goods, clothing, intact materials in demolition debris, building materials, business supplies and equipment, lighting fixtures, and any manufactured item or naturally occurring object that can be repaired or used again as is), paper, metals, glass, textiles, plastics (including plastic cases of consumer goods such as telephones or electronic equipment, film, and tires), plant debris, putrescibles (including animal, fruit, and vegetable debris, cooked food, manure, offal, and sewage sludge), wood, ceramics (including rock, tile, china, brick, concrete, plaster, and asphalt), soils, and chemicals (including acids, bases, solvents, fuels, lubricating oils, and medicines).

The list seemed fairly comprehensive, but there was always something, at least in my garbage, that seemed to lie outside the bounds of reuse. When I asked Knapp if he thought Zero Waste was truly possible in its strictest form, he barked, "Yes! But you've got to have hundreds of categories—that's the point of recycling. Garbage is just one category. Urban Ore has twenty-eight categories of doors. Nonferrous metal has thirty-five to forty categories. We landfill less than five percent of the five thousand tons we process each year."

"What's in that percentage?"

"It's painted wood and dust and dirt. Some film plastics."

I was surprised. "Does that include all those eight-track cassettes and old videos on, like, how to refinish a chair?"

"People buy and watch those, believe it or not. Or they make art out of them." I held my tongue, wondering how much art made of eight-track cassettes the public could stomach.

"What's your own trash like, at home?" I asked.

"My garbage is about six pounds a week, for two people. It's milk cartons and plastics. I bring the containers to a drop-off center. I don't use the provided recycling system where I live, in Richmond. They mix it all up. Commingled is contaminated. MRFs give you grossly contaminated stuff. Glass producers don't want sand and grit in it, and paper recyclers don't want glass in their stuff."

I was still getting used to the idea that everything in my kitchen trash bin would fit into a master category. "What about rubber bands?" I asked. Every day I got one from the delivered newspaper, one from the bundled mail, and sometimes another from the stems of fresh produce. I refused to throw them out, but I worried my collection would someday rival that of the Keishk brothers, who had built a 2,700-pound ball of rubber bands right here in San Francisco. (The Keishks bought their rubber bands in bulk from a company in Ohio.)

Knapp said, "There's a company in Berkeley that makes interesting personal goods from rubber—stuff like handbags. But I can compost them. They're not latex but synthetic, and they will break down." (One day I had an epiphany: I gave all my rubber bands back to the letter carrier, who was happy to reuse them.)

I told Knapp about the time I'd spent looking at San Francisco's MRF and the compost operation. He wasn't impressed. "Norcal is still just a big company making money off waste," he said. "There should *be* no waste." And the MRF? "A MRF is just another cover for a monopolistic operation. What I want to see is complex economic development—a quality operation. If you're going to achieve Zero Waste, you've got to use all the tools in the toolbox. You need less packaging, you need manufacturers taking it back, designing things that can be broken down or adapted. You need recycled-content laws. You need to provide an infrastructure for entrepreneurs to function."

* * *

Back in New York City, waste capital of the world, I pondered how much energy it took to push around San Francisco's carrot tops, pinot noir bottles, and sushi containers. The work provided jobs, and it probably kept some natural resources—nitrogen, silica, fossil fuels—in the ground. But I couldn't help thinking at times that we were, at great cost, shifting our messes from one place to another while a private waste management company toted up its earnings.

I believed that making new things from old things saved natural resources, including vast amounts of energy. (Even John Tierney acknowledged that recycling reduces pollution and energy use overall.) But I knew also that individual recycling efforts were puny compared to the larger world of waste. Of all the waste generated in the United States—including mining and agricultural waste, oil and gas waste, food processing residues, construction and demolition debris, hazardous waste, incinerator ash, cement-kiln dust, and other categories too rarefied to describe—municipal solid waste represented a mere 2 percent. Two percent!

Yet even when residents doubted the efficacy of recycling—and many New Yorkers did, thanks to the suspension and, of course, Tierney—they continued to go through the motions. Across the nation, more people recycle than vote. Recycling is a religion for some, and setting out bottles, cans, and newspapers is redemptive—a spiritual balm for our collective solid-waste guilt. We are powerless to replenish the ocean's fish stocks, to scour the chemicals from our rivers, to combat rising atmospheric temperatures, to halt the spread of exotic weeds and the global decline in biodiversity, but we can, by god, continue to sort our garbage, to make our offerings at the curb.

In a 1996 *In These Times* article called "Pavlov's Pack Rats," Joel Bleifuss wrote, "In many ways, recycling can be seen as a perverse form of penance in which individual recyclers absolve themselves from participating in an environmentally destructive culture." Bleifuss called the penance perverse because it wasn't individuals producing garbage but corporations—the soda bottlers, the catalog companies, the toy manufacturers, and the food packagers.

I didn't completely buy Bleifuss's argument. No one had held a gun to my head and forced me to buy a plastic wind-up fish for my daughter. I did choose naked green peppers at the co-op over plastic-wrapped peppers at the supermarket, but the choices weren't always so easy. Who made a wind-up fish that biodegraded? Or sold printer cartridges without a blister pack? Too often, our only choices were between one overpackaged, poorly made product and another. I came down somewhere in the middle of the argument: individuals could not shirk responsibility for the waste that passed through their hands, but neither were producers blameless. I had begun my travels in trash writing letters to manufacturers about their excessive packaging, single-use products, and Frankenstein designs. None had responded. I knew it would take laws (like landfill bans on single-use beverage containers and mandates for recycled content), sharp penalties, and massively collective consumer action to induce change. Still, no matter how trivial my kitchen recycling operation might appear, I considered it a moral act. It reminded me of the connection between my daily life and the natural world, from which every bit of the stuff arrayed about me had come.

Chapter Fourteen

The Ecological Citizen

"Sorting garbage is the ultimate Zen experience of our society," the Garbage Project's William Rathje has said, "because you feel it, you smell it, you see it, you record it; you are in tactile intimacy with [it]. Sometime or other, everybody ought to sort garbage."

I had been pawing through my garbage for nearly a year, sorting it into piles bound for the landfill, for recyclers, for my compost bin, and for hazardous-waste quarantines. I was sick of the mess in my kitchen, the weighing of bottles and cans, the bags. And yet I couldn't quite let my garbage go. The budding social theorists in Robin Nagle's New York University class called garbage "unmarked," as opposed to marked, which meant it was not studied and was hardly remarked upon. Now that I was done quantifying my waste, would it once again become unmarked to me? Already I felt my attention slipping. Just as I had quit noticing what type of strollers were de rigueur once I stopped perambulating with an infant, would I soon fail to notice if a paper receipt landed in the trash can instead of in the recycling? It made me a little sad to think of mindfulness fading.

And so I retired from weighing my garbage but continued to sort it into piles. It was a habit, and I'd feel guilty if I quit. I knew too much about where my refuse was headed, about its potential impacts on people and other living things. The conservationist Aldo Leopold once wrote, in his essay "Round River," "One of the penalties of an ecological education is that one lives alone in a world of wounds." I'd always found the quote a little grandiose, but it came to mind whenever I was tempted to put something divertible into the regular trash. I wasn't convinced my compulsive sorting was doing much good, but it made me feel less bad about so many other things.

Since the beginning of July, the Hugo Neu Corporation had been taking, in addition to household metal, the city's gable-top containers and narrow-necked plastic bottles numbered 1 and 2. I saved plastic bags and large-mouthed plastic numbered 2, 4, and 5 for the food co-op; tied up paper for the curbside collection; put organics in the compost pile; set beer bottles on the street for Willy; saved batteries and other household hazardous waste in a basket destined for the drop-off center; and left everything else—including glass, for which a market had yet to be found—for John Sullivan and his new partner, Mike Perrani (Murphy had finally retired). Unfortunately, the city had dialed back recycling pickups to once every other week (and the co-op accepted recyclables on the same schedule), so the piles grew, and my kitchen was more like a MRF than ever.

The compost wasn't in such great shape, either. My neighbors on the first floor had taken some "rich, dark, soil-like material" from the bottom, but the bin was still pretty full of recognizable food scraps. One morning I went downstairs to collect some dirt for my houseplants: I'd doused half of them with NYOFCo's pelletized biosolids a few months earlier, and the hopped-up fertilizer had nearly killed them. I attacked the top layer of the compost with the potato fork and immediately recoiled in horror. My tines had upturned a writhing mass of worms. And these were no gardener's friend, the slim red *Eisenia foetida*. They looked more like bloated maggots, ribbed, white, and about an inch long. The worms, or larvae, made me think twice about my plans for that ultimate piece of garbage, my own dead body. Cremation is energy intensive and polluting, especially if you have mercury in your teeth; and conventional burial pollutes groundwater with embalming fluids. I'd always thought I'd like to be buried, au naturel, in a wood near a body of water, something like Walden Pond, and let the worms do their thing. Now I had second thoughts.

Deeply disturbed by the pasty grubs, I reburied them and in a zombielike trance went back upstairs.

When I finally sat down to crunch my kitchen garbage numbers, I found that two adults and one child had in ten months sent to the dump an average of 4.65 pounds of trash a week, which was less than one nineteenth of the national average (using the EPA's per person figure). Over that same period I'd theoretically diverted 680 pounds of metal, glass, plastic, and paper (glass collection was still suspended, so after weighing all those jars and bottles, I regretfully placed them back in the trash), and tumbled 221 pounds of green material into my compost bin. Had all 1,100 pounds gone to Bethlehem on my own private truck, it would have cost me fifty-seven dollars to tip.

I was wondering how other Americans managed to throw away 4.3 pounds of trash a day until I realized that I hadn't done a major household purge during the course of my study. (I hadn't included bathroom or bedroom trash in my tallies either, but they wouldn't have added more than a few ounces a week.) So I went through a burst of housecleaning and placed a lot of unwanted stuff on the sidewalk. Passersby adopted two tote bags, a set of bowls, two dozen children's books, and two working printers in less than five hours. But there was still some twenty pounds of unsalvageable household chattel in my trash can: a stroller, a mop, a lamp, a Magna Doodle, four rust-stained cotton curtains. Was that enough to seriously spike my daily average? No. But if I didn't live in a neighborhood where I could leave stuff on the sidewalk for others, didn't have access to curbside recycling and a compost bin, and had to throw out a major piece of furniture, I'd be right up there with the rest of the nation. The exercise helped me realize I was lucky to live where I did, and that it was easy to be good for a month or two. But after that it got harder. My family members had thrown out, over the course of ten mindful months, an average of three and a half ounces a day. It sounded good compared to the national average, but still: we were no Armantrouts.

I pored over my data, which included the name of every item in my trash can. What would an anthropologist make of my diet, based on these dregs? Because I'd diverted organics and recyclables, my garbage—unlike that of my grandparents fifty years ago—wasn't really representative. It appeared as though I ate no ice cream, for one thing (I put the cartons on the paper recycling pile). And drank no beer (but plenty of wine, which came in a nondeposit bottle). Once I started bringing plastic bags to the food co-op, my garbage implied that I didn't eat bread, nuts, rice, or raisins. All evidence of my healthy diet had been diverted to the compost pile or American Ecoboard, in Farmingdale, Long Island.

An anthropologist would surmise a child lived in the house (because of the craft projects and the occasional appearance of small, stained clothing), that a great crash had occurred in the kitchen (eight smashed dinner plates and four cracked bowls), that the family had weathered an infestation of pantry moths (full traps), and that allergies had plagued an adult (foil from Sudafed tablets).

My garbage log revealed that just because you're a child doesn't necessarily mean you generate less trash (by volume, not weight). Lucy produced toy waste, arts-and-crafts waste, clothing waste, and, because she had the usual dietary peculiarities of a four-year-old, far more food and food-packaging waste than her parents did. A lot of Lucy's remains were evidence of pleasure—and that heartened me. But just as each additional person on the planet further strained our dwindling natural resources, so did that person's garbage further strain the earth's capacity to benignly receive it.

Like anyone with a garbage obsession, I'd begun to pay closer attention, over the previous year, to how grocery items were packaged. If their wrappings were unavoidably "bad," I could at least buy a larger size, which reduced the amount of packaging per serving (or squirt, in the case of cleaning products). I got better at refusing plastic grocery sacks, until the supply under my kitchen sink dwindled from the size of a compacted laundry load to nothing. I had almost entirely quit using disposables: I went through less than one roll of paper towels the entire year, and Lucy kept asking if we could please bring out the special napkins—the paper ones—when company came.

I'd also begun to act differently around my garbage. If it started to rain, I tried to run downstairs and cover the cans. I tied up my cardboard the way Jorge and Jack wanted, and I taped the ends of anything sharp. I positioned the trash cans on the sidewalk in a gap between parked cars. I cultivated a respect for the men and women who almost invisibly whisked it all away. This is probably one of the hardest places in the world to be a sanitation worker. New York has more people packed tighter, and more garbage with fewer places to put it, than any other city in the nation. The streets are narrow, the languages diverse, and the right to double-park appears to be god-given. Walking down the street now, I'd hear the distinctive knock of a packer truck's diesel engine and turn to see if it belonged to the Brooklyn 6. Perhaps one of the most important things I'd learned in the past year was the names of the people who took away my trash.

At the recycling roundtable many months earlier, I had met Samantha MacBride, who worked at the Department of Sanitation's Bureau of Waste Prevention, Reuse and Recycling. "Recycling isn't saving the earth," MacBride had said to me that November afternoon, her face arranged in a frown. "Just so long as you know that. There are very few environmental benefits to recycling." A PhD candidate in New York University's Department of Sociology, MacBride had written a paper that challenges the value of consumer recycling. Such programs, she wrote, redirect "the focus of environmental concern away from [the] material unsustainability of the current economic system, instead turning it inward on the self." Individual recycling was not only unhelpful, she believed (because it was such a slim fraction of the overall waste stream), it was also a shining example of how individual goodwill had been perverted by capitalist goals. Household recyclers, she wrote, were "simultaneously uninformed and concerned about ecological problems, as well as enthusiastic and active in largely meaningless solutions."

Surely my ersatz kitchen MRF would have amused MacBride. By trying to shrink my garbage footprint, I was—in her paradigm—abetting a bankrupt system by doing what the government, educators, and environmentalists (who increasingly partnered with corporations) told me to do. Recycling merely made it easier for individuals to keep consuming and to keep discarding. It also gave waste hauling companies who ran recycling programs an opportunity to look as though they cared. (And to make more money.)

Perhaps it would have been more radical for me to opt *out* of recycling: to throw everything into one sack and set it on the curb. A lot of people did that, though absent a political philosophy. In Edward Abbey's novel *The Monkey Wrench Gang,* George W. Hayduke chucks empty Schlitz cans out of his Jeep onto the highways of the desert Southwest. "Why the fuck shouldn't I throw fucking beer cans along the highways?" he asks his friend Seldom Seen Smith.

"Hell," Smith says, "I do it too. Any road I wasn't consulted about that I don't like, I litter. It's my religion."

"Right," Hayduke says. "Litter the shit out of them."

MacBride's critique of household recycling sprang from the EPA's mind-boggling statistic, that municipal solid waste constitutes just 2 percent of the nation's waste, with the remainder (some 12 billion tons) nonhazardous industrial waste (NIW), plus mining, agricultural, and hazardous waste. (About 75 percent of NIW is in liquid form. Think hog and cattle lagoons. But even discounting this watery portion, NIW still outweighs MSW eleven to one.) The average American knows nothing about NIW. It is, in Robin Nagle's word, "unmarked." States don't track its composition or its amounts, and neither the federal nor the state governments regulate its disposal (thanks to industry's campaign contributions and heavy lobbying, suggested MacBride; the less the public understood the impacts of consumption, the less likely it would be to challenge producer systems).

Adam I. Davis, a founder of the Natural Strategies consulting firm and a former compost and fuel programs coordinator for Waste Management of North America, connected each ton of municipal solid waste with approximately sixty-one tons of waste from primary extractive industry and manufacturing. "For the roughly 210 million tons of MSW the US generates each year," he wrote in 1998, "we generate an additional 7.6 billion tons of industrial waste, and 1.5 billion tons of mining waste, 3.2 billion tons of oil and gas, electric utility and cement kiln wastes and .5 billion tons of metal processing waste."

The disparity between my personal waste and the waste it took to produce my waste shocked me, but it didn't mean that the tiny fraction in my cans was inconsequential. I had seen the garbage piles and the trucks they rode around on, and heard the anger and frustration of citizens who live at the margins where garbage settles. Two hundred thirty-two million tons of municipal solid waste a year, the EPA's national figure for 2003, isn't a small pile. But the more important point is that the 2 percent is not unrelated to the 98 percent, which has everything to do with the back end of our upwardly mobile lifestyles. Remember William McDonough: "What most people see in their garbage cans is just the tip of a material iceberg: the product itself contains on average only 5 percent of the raw materials involved in the process of making and delivering it." And remember Paul Hawken: for every 100 pounds of product that's made, 3,200 pounds of waste are generated.

No one's numbers agreed, but the multiplier effect nonetheless kept me working to return steel and paper, if not always plastic and glass, to manufacturers. If a single barrel of waste on my Park Slope curb was indicative of thirty-two barrels of manufacturing waste, then by halving my garbage I could eliminate sixteen barrels up the line. While diverting items to the recycling bin was good (it would avoid some use of virgin materials), not buying quite so many things in the first place was far better.

What would happen if we slowed our pace of buying, if we kept our well-made furniture, appliances, and cars for life, or even twice as long as we did now? Would the economy grind to a halt, throwing millions of workers onto the welfare rolls? Advocates of green design imagine a sunny future of high employment. It would take a lot of people to refashion everything using only biological and technical nutrients. (According to the Institute for Local Self-Reliance, sorting and processing recyclables—of which everything in a green-design world would be made—employs ten times more people than landfilling or incineration, on a per-ton basis.)

A rising tide of fed-up consumers might be able to chip away at the treadmill of production, demand that our government address the disparity between producer and consumer wastes, and offer manufacturers an incentive to produce efficient, reusable products and packaging, but on my own I felt helpless to do anything about the 98 percent. And so I continued recycling and watching what I bought. It was something I could manage. The idea of Zero Waste inspired me because it was linked with serious design and manufacturing changes. Consumer goods had to be less toxic, designed for recycling, and easily returned to their makers. I was more than willing to consume better, if manufacturers would only agree to produce better. Until that time, we have the hierarchy of waste: reduce consumption, reuse consumer goods, and recycle and compost the rest like crazy.

I hoped that some day we would look back on the piles of garbage on our curbs and the compressed bales of paper, metal, glass, and plastic in the transfer stations, and shake our heads at the primitiveness of it all. We were in an awkward in-between stage now: we generated a lot of waste, we worried about it, and we were motivated to take individual action, no matter how misguided. I wanted to believe this moment in garbage history was a blip along our route to sustainability.

Were we on the cusp of some brilliant technological breakthrough? I doubted we'd find something revolutionary to do with all our garbage, like cooking it into inert cubes that we sunk in the ocean or blasting it into outer space (an option put forth by a surprising number of san men). Furthermore, technological fixes might legitimate the generation of waste at our current rate, and they had a way of magnifying our effect on the planet. "All of our current environmental problems are unanticipated harmful consequences of our existing technology," the evolutionary biologist Jared Diamond has written. "There is no basis for believing that technology will miraculously stop causing new and unanticipated problems while it is solving the problems that it previously produced."

In the days before e-mail, inventors and entrepreneurs used to burst into the DSNY's headquarters with "solutions" for the city's trash "problems" (the widespread use of computers considerably cut foot traffic). "I had one guy who wanted each truck to burn its own load of garbage as its own fuel source," said Vito Turso, deputy commissioner for public affairs. "That would mean two thousand mobile incinerators traveling and belching around New York City each day." Then there was the gentleman who proposed stacking tires, along with construction and demolition debris, on the floor of New York Harbor. When the pile rose above high water, he'd build an incinerator atop his brand-new "Recap Island." Another engineer envisioned trash barrels positioned over a vast underground vacuum, which would suck garbage into a waste disposal facility. "I like to call that one Chutes and Litter," Turso said, laughing. "People will continue to come up with these ideas as long as we continue to come up with garbage. You try and you try and you try, and eventually one of them might work out." But as New York's former commissioner of sanitation Norman Steisel once said, "In the end, the garbage will win."

New York City is famously resistant to change, but there are plenty of other communities and private companies willing to experiment with waste. There is the Nantucket digester, for example, the green roofers, the recyclers of gray water, the machine in Santa Clarita, California, that chops up dirty diapers, sends the waste to sewers, dries out the wood pulp, and pelletizes the plastic, which is given to the local MRF. Cargill is turning refined corn sugar into plastic that biodegrades in a commercial composting operation; Changing World Technologies is building a plant in Missouri that, through thermal conversion, can transform old tires, plastic bottles, sewage, or slaughterhouse scraps into fuel oil. New stuff is coming along all the time—collection technologies that could eliminate all household sorting; MRFs that consume enough bottles, cans, and paper to stabilize market prices and introduce economies of scale.

While waiting for recycling technology to advance, I took heart from the principles of source reduction and the growth of Craigslist and "freecycling" Web sites, which help folks move unwanted goods laterally through their communities. I'd become addicted to cruising these virtual communities. Some opened a window onto private lives ("Free guinea pigs: new boyfriend is allergic"), some seemed more like sociological research into what sort of person answered these ads ("Free coal!"), and some were just funny ("25 crappy futons—not in great condition but they're also not disgusting or infested so they're definitely usable.")

After a year steeped in garbage issues, I know that finding a place to put our waste isn't the problem. There is plenty of landfill space and incinerator capacity out there. But transporting waste and tipping it are becoming increasingly expensive. New York's sanitation budget for 2003 approached $1 billion. And no matter how sophisticated their design, landfills and waste-to-energy plants are hardly environmentally benign. They leak and they spill, and their poisons are slowly accumulating in the creatures with which we share this planet. Landfills can't be safely reused for hundreds of years, if then, and incinerated materials are lost to reuse forever.

If we have a garbage problem, it is that landfills and incinerators make it too easy to get rid of things. Burying or burning waste only spurs more resource extraction to make more products. Our trash cans, I believe, ought to make us think: not about holes in the ground and barrels of oil saved by recycling, but about the enormous amount of material and energy that goes into the stuff we use for an instant and then discard. Garbage should worry us. It should prod us. We don't need better ways to get rid of things. We need to *not* get rid of things, either by keeping them cycling through the system or not designing and desiring them in the first place.

My interest in garbage had been piqued on Earth Day, when volunteers plucked trash from the Gowanus Canal. I had doubted the value of the exercise and had politely said no to a dip net. A year and a half passed, and a friend asked me to participate in a beach cleanup organized by International Coastal Cleanup in the Fort Tilden section of Gateway National Recreation Area, in Queens.

By now I knew too much about our combined sewer overflows to think that cleaning this sandy strip would do any good. The next big rain would bring a load of trash from the sewer pipes, and a sunny day would bring beachgoers who littered. Moreover, I knew that several of International Coastal Cleanup's sponsors—including Dow Chemical, the Coca-Cola Company, and the American Plastics Council—had spent tens of millions of dollars fighting bottle bills, which made their support of litter campaigns more than a little hypocritical. I saw the symbolic value of litter collection, to be sure. And frolicking on a clean beach, canal, or creek would certainly be more pleasant than picnicking among someone else's plastic wrappers or paddling a fetid inlet. But community cleanups had an insidious way of shifting responsibility from those who knowingly produced the mess to individuals willing to pick up the slack. Still, it was a sparkling fall day, and the lure of the beach was stronger than my misgivings. I buckled Lucy into her car seat.

At the meeting place, volunteers handed out plastic garbage sacks and gave Lucy a cardboard fan—it read "I'm a fan of clean water"—that I knew would shortly be in my recycling pile (except for its wooden handle, which I'd save for kindling). We were also offered small plastic business card holders embossed with a sponsor's name. Accidentally drop it in the street, and the odds were good it would end up right here, on the beach.

As it turned out, plastics were the number-one category of litter we collected that day: plastic caps, plastic bottles, plastic wrappers from food. A lot of this came from beachgoers, but most of it had washed off the city streets and into the waterways after a big rain. In thirty minutes, I picked up ten cigarette butts (Philip Morris was one of today's sponsors). In 2002, International Coastal Cleanup volunteers in 117 countries collected 1.8 million cigarette butts and cigar tips, representing 31 percent of all trash items. By design, cigarette filters trap toxic chemicals before they enter the smoker's body, but tossed on the ground they leach toxins into the environment. Cigarette butts have been found in the stomachs of fish, birds, whales, and other marine creatures that mistake them for food. A New York City Parks Department gardener, writing in *Slate,* suggested that filters be made of something organic—she suggested chicken poop—which would at least enhance city flowerbeds, where so many smokers insisted on tossing their butts.

Over the course of the morning, two dozen volunteers on our short stretch of beach filled eighty bags with litter while noting on data cards their exact contents. If the others' bags were anything like mine, they contained a fair share of recyclable cans and deposit bottles. I called later to find out what happened with the trash bags and learned they had gone straight to a putrescible waste transfer station, unopened.

Throughout New York State, more than ten thousand people had volunteered to clean 385 miles of waterway. While I was picking through the sands of Fort Tilden, volunteers on the Gowanus were collecting enough junk to fill more than a hundred bags. In 2002, International Coastal Cleanup had collected more than 8.2 million pounds of refuse, including 16,554 condoms, 16,144 tampons, 22,759 diapers, 255,972 straws and stirrers, 335,070 plastic bags, 347,137 glass bottles, 360,104 aluminum cans, 423,820 plastic bottles, and 675,360 food wrappers.

I had been quick to ridicule the beach cleanup, but there was an ethical component to the activity that I couldn't deny. The environmental philosopher Andrew Light believes that living in a city entails a form of political citizenship as well as a form of ecological citizenship, which he loosely defines as "a ground of moral and political environmental responsibility for one's duties to the human and natural communities one inhabits and interacts with." Citizens engaged in a relationship with the land around them are less likely to harm it or to let it be harmed. Whether a spartina meadow or a cleaner beach, the end product doesn't matter. What is important in Light's scheme is the values created in that practice—a relationship with nature, a community committed to the protection of a local environment. "It is a restoration not only of nature but also of the human cultural relationship with nature." I wasn't sure how to characterize my relationship with the Fort Tilden beach, but I did feel like coming back here soon, if only to see how long the clean stuck.

A week after my introduction to ecological citizenship, I drove out to Barren Island to meet Andy Bernick, an ornithologist who had helped me gather information on bird counts at Fresh Kills. I wasn't looking for birds this time, though Bernick had brought his spotting scope. I was here to see garbage.

One of the largest islands in Jamaica Bay, Barren Island had once comprised about thirty acres of sandy upland and another seventy acres of salt marsh. Used by Brooklyn farmers to grow salt hay and graze horses, the island was accessible from the mainland only at low tide and remained uninhabited until the end of the eighteenth century. In the late 1850s, two fertilizer factories opened and began processing dead horses and other animals shipped from New York. According to National Park Service historian Frederick Black, "Between 1859 and 1934, perhaps as many as twenty-six companies [manufacturing establishments] had facilities on the island, although no more than seven or eight at any one time." The factories produced fish oil from menhaden, which migrated in huge schools along the Atlantic coast and were captured by fishermen in seines. The oil was used in leather tanning and paint mixing; the fish scraps were used as fertilizer.

Menhaden populations eventually declined, but not before scheming politicians arranged to ship city refuse to the rendering factories of Barren Island. In the busy season, resident laborers unloaded seven or eight scows a day, a total of three thousand tons of refuse from Manhattan, Brooklyn, and the Bronx, in addition to the daily horse boat, which held as many as fifty dead horses, plus cows, cats, and dogs. Workers picked through the garbage for valuables, then boiled or steamed the rest in fifteen-foot-high steel cylinders. By 1860, writes Benjamin Miller in *Fat of the Land,* the island had "the largest concentration of offal industries in the world," producing fifty thousand tons of oils and tens of thousands of tons of grease, fertilizer, and other products (bone black, hides, iron, and tin) worth more than $10 million a year.

By the turn of the nineteenth century, Barren Island was home to some fifteen hundred people, with their attendant butcheries, bakeries, churches, and saloons. It was not a pleasant place to live. The raw garbage stank, and the cooked garbage was worse. The stench, according to the Brooklyn Health Department, had "a range, flatness of trajectory and penetration equal to that of our modern coast defense rifles."

Politics and economics closed manufacturing plants through the 1920s, and by 1935, the single remaining factory on Barren Island was dismantled. Starting in 1909, the area north of the island was filled in with city refuse, and numerous small islands were connected for the construction of houses and, eventually, Floyd Bennett Field, the city's first municipal airport. In 1936, Robert Moses evicted the last residents of Barren Island, which was no longer an island, in order to build his Marine Park Bridge, which brought motorists to the Rockaway Peninsula and Fort Tilden, where I'd collected the flotsam of more recent adventures. Once again, the roads had brought the people, and the people had brought their trash, mostly the kind of stuff—individually wrapped snacks and disposable beverage containers—that didn't exist when people weren't quite so mobile.

I'd visited Barren Island only once previously, long before I understood its place in New York's garbage history. I had come here from Prospect Park in search of a pastoral experience: I went camping. Eighty years earlier, Barren Island's schoolmistress had made the reverse trek, bringing her young charges from their soot-darkened confines to Prospect Park. What the children did there is lost to history. I spent half my time on Barren Island in a tent and was nearly run over in the middle of the night by teenagers rampaging in a Jeep across the former salt marsh.

A City University of New York graduate student, Andy Bernick had rounded cheeks and short, glossy black hair. He wore black-rimmed oval eyeglasses, a checked shirt, jeans, and sneakers. Bernick had almost superhuman patience. As we walked toward the beach down a mowed path, I couldn't help peppering him with excited questions, often interrupting his quiet explanations. He didn't seem to mind; he was excited to be here, too.

We came out onto the beach along Dead Horse Bay, and Bernick said, "Ooh, this is good." He meant the tide: it was superlow, and the exposed mudflats were thirty to forty feet wide. Before the rendering plants went up, this now-gentle curve of beach was called Dooley's Bay, which was actually the outwash of a large creek that had cut Barren Island almost in half. Now, with the water so low, I could barely make out the bay's shallow contours.

The first thing I noticed on the beach were the rotted pilings that marched out into a channel—the remnants of docks built to unload scows. I looked down and saw the usual New York beach litter: cups, chip bags, plastic bottles and caps, broken glass. "I guess the Coastal Cleanup didn't make it here last week," I said. Then we walked a few meters toward the water and, bam, it hit me. The entire beach appeared to be tiled with glass, with brown patent-medicine bottles and ceramics, dainty teacups and bottles blue, green, and clear. Mixed in with the shells of horseshoe crabs, sea lettuce, periwinkles, crab legs, and condensed marsh muck were rubber tires, jug tops, shoe heels, and soda bottles with typefaces long abandoned.

The sea lettuce, which grew in plate-sized sheets, was a product of nitrification: too much nitrogen came from the three sewage treatment plants that discharged into the bay (a fourth, the nearby Rockaway plant, discharged into the ocean, though currents likely swept effluent in here, too). Overstimulated by the fertilizer, the lettuce ran rampant. It consumed oxygen needed by other organisms and it smothered other aquatic life. The city was under court order to reduce the amount of nitrogen it discharged into waterways, and both New York and Connecticut had signed binding commitments with the EPA to reduce nitrogen in Long Island Sound, where another four city plants discharged. The goal was a 58.5 percent decrease, by 2014, over 1990 levels.

Bernick set up his spotting scope, just to see what he could see, and I walked the tide line, stopping to pick things up and marvel at the processes of nature and the shortsightedness of man. Jamaica Bay was slowly rising, and its islands were subsiding. No single theory explained the shrinkage, but scientists suspected that development had reduced the flow of sediment feeding the marshes, and sea lettuce was smothering them. The trash heaped under the sand was slowly being exposed by the action of winds and tide. "Every time I come down here it's different," Bernick said. "The beach looks different, and the garbage is different." The layers most recently dumped were the first to be revealed; as time passed the oldest material would come to the fore. We guessed that today's snapshot—a glass pipe, an old spool, vinegar and ale bottles, chicken bones, crockery, an octagonal blue Milk of Magnesia bottle, a brittle plastic doll—had been buried in the forties or fifties. The reduction plants of a century past had left very little behind. Still, it was easy to find horse bones on this beach.

Folding up his scope, Bernick and I rounded the shoulder of the bay. Here, the beach rose to a meter-high bank topped by phragmites and spilling some creamy white guts—a tangle of thick nylons filled with sand. The stockings were reinforced for garters; some had seams up the back. The movement of water had twisted the mass into a soft Mummenschanz sculpture. In some areas, the junk was segregated by type: a section of white cold-cream jars, a section of ketchup bottles, a section devoted to crushed bottle necks, a region of white ceramic fixtures interspersed with metal machine parts rusted thin. Bernick and I walked with our eyes downcast, lost in thought. Now and then we tossed things at each other—a doll arm, an Empire State license plate. Everywhere, we found leather shoes: men's boots, women's pumps, children's lace-ups. Twisted and bent, they invoked in me an overwhelming sense of poignancy. They all seemed to have been worn to death, and the people who'd worn them were long gone.

Most of the refuse on the beach was glass. It glittered in the afternoon sun and its broken shards tinkled delightfully at every crest of wavelet upon the shore. Why was there so much glass? Because glass is what remains. An inorganic, nonbiodegradable solid, it didn't float "away," like plastic; it didn't have value for recyclers or scavengers; and it didn't break down into tiny pieces, on its own, over human-scale time. Just as the modern city had found no outlet for its Everests of shattered household glass, neither had the waste brokers of Barren Island.

I collected an ale bottle I thought Lucy would like and some shards of pottery hand-painted in Italy, but she wouldn't immediately latch on to them, and eventually I'd put them in my trash. If the city recycled and made money from glass, my idle scavenging—which was illegal, as we were on National Park Service property—would have been called poaching.

Some manufacturers and processors, like Hugo Neu, were moving toward dumping their separated wastes in "monofills" that they'd maintain until technology evolved and the waste became valuable. It didn't seem that glass was headed in that direction. Even die-hard environmentalists admitted that collecting and processing glass could consume more energy than it saved. They said there was no shortage of sand, the principal ingredient in glass, and that it wasn't toxic in a landfill. Left alone, the bottles on this beach would eventually return to sand. The thought gave me a creepy *Planet of the Apes* kind of feeling. Barren Island would be under several feet of water by then.

But what about in fifty or a hundred years: what will the garbage landscape look like then? Already, the stuff we set on the curb is circling back to bite us. We burn our electronic waste, and its chemical fallout shows up in the breast milk of Eskimos and in the flesh of animals we eat. We bury our household waste, and poisons rise into our air and leach into our waterways. We can recycle and compost as much as we want, but if the total waste stream continues to grow—and it *is* growing, whether in places where recycling is on steroids, like Seattle, or in places where recycling is anemic, like the entire state of Mississippi—we'll never escape our own mess. For better or worse, consumers and producers respond to economic arguments: if we don't wake up and make the connection between our economy and the environment (which provides the resources to make all our stuff), the planet will eventually do it for us. And it won't be pretty.

When I got home from my Barren Island outing, I learned that the Coastal Cleanup folks *had* descended on Dead Horse Bay, filling their plastic sacks with today's seaborne litter and yesterday's heavier fill. The bags went onto a truck and joined the parade of refuse heading out of state. There was nothing more personal and local, I thought, and nothing more inadvertently global than an individual's garbage. Three hundred years from now, I imagined someone excavating a Pennsylvania hollow, perhaps in search of raw materials no longer abundant in nature, and pondering how an ale bottle, filled and refilled in nineteenth-century Brooklyn, had made it so far from home.

Acknowledgments

Among the many people who helped me with this project, I'd like to single out Daniel Katz, Benjamin Miller, Robin Nagle, and Lisa Reed for their early leads and inspiration; Bob Besso, Dennis Diggins, Virali Gokaldis, Allen Hershkowitz, Robert Lange, G. Fred Lee, Eve Martinez, Peter Scorziello, Steve Shinn, and Nick Themelis for fielding endless waste-related questions; Phil Heckler for his help with the pipes; Carl Alderson for misreading the tide chart; and Tony Israel for trying to save the earth. I'd like to offer a 2.9-billion-cubic-yard thank-you to my editor, Geoff Shandler, who saw through the muck; to Liz Nagle, whose comments were always spot-on; to Betsy Uhrig, copyeditor extraordinaire; and to my agent, Heather Schroder, for her support and guidance. I'm especially grateful to my husband, Peter Kreutzer, for closely reading and commenting on every version of the manuscript, and to Lucy, who almost always said yes to an outing.

A Note on Sources

To research this book I relied on many books, reports, and articles. Among the most valuable were:

Bronx Ecology: Blueprint for a New Environmentalism, by Allen Hershkovitz, Island Press, 2002.

The Consumer's Guide to Effective Environmental Choices: Practical Advice from the Union of Concerned Scientists, by Michael Brower and Warren Leon, Three Rivers Press, 1999.

Cradle to Cradle: Remaking the Way We Make Things, by William McDonough and Michael Braungart, North Point Press, 2002.

Fat of the Land: A History of Garbage in New York; The Last 200 Years, by Benjamin Miller, Four Walls Eight Windows, NY, 2000.

The Meadowlands, by Robert Sullivan, Scribner, 1998.

My Life in Garbology, by A. J. Weberman, Stonehill, 1980.

Natural Capitalism: Creating the Next Industrial Revolution, by Paul Hawken, Amory Lovins, L. Hunter Lovins, Little, Brown, 1999.

The Road to San Giovanni, by Italo Calvino, Pantheon Books, NY, 1993.

Rubbish!: The Archaeology of Garbage, by William L. Rathje and Cullen Murphy, University of Arizona Press, 2001.

Takedown: The Fall of the Last Mafia Empire, by Rick Cowan, Random House, 2002.

Toxic Sludge Is Good for You!, by John Stauber and Sheldon Rampton, Common Courage Press, 1995.

Underworld, by Don DeLillo, Scribner, 1997.

Up in the Old Hotel, by Joseph Mitchell, Pantheon, 1992.

Waste and Want: A Social History of Trash, by Susan Strasser, Metropolitan Books, NY, 1999.

Biocycle magazine's "State of Garbage in America," 2004.

"The Case for Caution," Cornell Waste Management Institute, 1997.

"Exporting Harm: The High-Tech Trashing of Asia," prepared by the Basel Action Network and the Silicon Valley Toxics Coalition, 2002.

"A Gospel According to the Earth," by Jack Hitt, *Harper's*, July 2003.

"Life After Fresh Kills: Moving Beyond New York City's Current Waste Management Plan," a joint research project of Columbia University's Earth Institute, Earth Engineering Center, and the Urban Habitat Project at the Center for Urban Research and Policy of Columbia's School of International and Public Affairs, December 1, 2001.

"No Room to Move: The City's Impending Solid Waste Crisis," City of New York, Office of the Comptroller, Office of Policy Management, October 2004.

"Overview of Subtitle D Landfill Design, Operation, Closure and Post-closure Care Relative to Providing Public Health and Environmental Protection for as Long as the Wastes in the Landfill Will Be a Threat," by G. Fred Lee, PhD, PE, DEE, and Anne Jones-Lee, PhD, January 2004.

"Pavlov's Pack Rats," by Joel Bleifuss, *In These Times*, November 13, 1996.

"Recycling Is Garbage," by John Tierney, *New York Times Magazine*, June 30, 1996.

"Recycling Reconsidered: Producer vs. Consumer Wastes in the United States," unpublished paper by Samantha MacBride, October 2001.

"Recycling Returns: Ten Reforms for Making New York City's Recycling Program More Cost-Effective," by Mark A. Izeman and Virali Gokaldis, Natural Resources Defense Council, April 2004.

"Report of the Berkeley Plastics Task Force," by Richard Lindsay Stover, Kathy Evans, Karen Pickett, April 1996.

"The Urban Blind Spot in Environmental Ethics," by Andrew Light, *Environmental Politics*, Spring 2001.

"Urban Residential Refuse Composition and Generation Rates for the 20th Century," by Daniel C. Walsh, *Environmental Science & Technology*, November 15, 2002.

"Wretched Refuse," by Keith Kloor, *City Limits*, November 2002.

Index

Abzug, Bella, 35
acid rain, 44, 79, 169
Addington, Bill, 214
Africa, 167, 193
African Americans, 43, 153, 213, 263–64
agriculture, 125, 213, 222, 225, 275
air quality, 24, 75, 77, 79
Alderson, Carl, 53–56, 60–62, 95, 97, 98
Allied Waste, 40, 41, 71, 72, 179–82, 188
aluminum, 145, 146, 155–56, 164, 172, 181, 256
American Ecoboard (Farmingdale, New York), 177–79, 186, 189, 190, 280
American Ref-Fuel, 77–79, 161
Apuzzi, Ed, 44, 45, 46–49, 72–73
archaeology, 83–84, 91–92, 240
Archibald, Judy, 75
Arizona, 128, 214, 266
  See also Garbage Project
Armantrout, Bob and Camille, 257–58, 280
Arthur Kill, 53, 58, 60, 95, 96, 99
Asia, 132, 153, 155, 167, 254
  See also China; India
asthma, 43, 44, 46, 213, 214, 219, 222
Australia, 254

bacteria, 90, 107, 110
  See also digesters, anaerobic food
Balan, Ludger K., 7–10
Balan, Mitsue, 9
Barber, Bill, 224
Barren Island (Jamaica Bay), 22, 289–94
Basel Convention (1992), 170
Basso, Dominick, 154–55
batteries, 158–60, 163, 164, 239, 268, 278
Benedetto, Sal, 72
Berkeley (California), 80, 189–90, 192, 201, 272–74
Bernick, Andy, 289–94
Berry, Wendell, 238
Besso, Bob, 254, 256, 258–61, 263–66, 268, 269
Bethlehem (Pennsylvania), 48, 53, 63–68, 73, 75–76, 81, 84
biodegradation, 90–92, 110, 122, 191, 286
bioreactors, 57–58, 91, 92
biosolids, 210–12, 221–23, 229, 272, 278
  See also sludge
birds, 9, 23, 51, 60, 67, 75, 88, 96, 110, 289
Black, Frederick, 289
Bleifuss, Joel, 275–76
Bloomberg, Michael, 14, 72
Blumenfeld, Jared, 263
bottle bills, 13, 20, 155–56, 182–86, 188, 287
bottles, 177, 194, 257–58, 262
  deposit, 13, 16, 36, 185–86, 251–53, 259, 278, 288
  refillable, 186, 252, 266, 294
  See also glass; plastic
Braungart, Michael, 122, 239, 240
Bronx (New York), 43, 115, 119, 137, 143, 212–13, 218–19
Bronx Ecology (Hershkowitz), 137
Brooklyn (New York), 3–10, 43, 132, 163–65
  Park Slope, 4, 5, 21, 27–35
  Red Hook, 23, 46, 199

Brooklyn (*cont.*)
Williamsburg, 40, 72, 80, 199
See also Owls Head sewage treatment plant
brownfields, 54, 64
Browning-Ferris Industries (BFI), 69–70, 71
Budnick, Nick, 35
Burrafato, John, 30–31
Bush, George W., 242

California, 11, 51, 172, 183, 222, 226–29, 254, 285
diversion rates in, 264–65
Integrated Waste Management Board of, 193, 264
landfills in, 51, 53, 58, 74, 268
recycling in, 128, 266
See also Berkeley; Los Angeles; San Francisco
Calvacca, Joey, 37
Calvino, Italo, 38–39
Campbell, Ed, 168
Canada, 41, 125, 138, 254, 260
carcinogens, 43, 44, 74, 93, 170, 191, 220, 236
Cardamone, Richard J., 70
cell phones, 166, 167, 174, 175
Chewonki Foundation (Maine), 201–2
children, 19, 21, 189
China, 153, 162, 169–70
Chinatown (San Francisco), 259–61
Christiansen, Kendall, 121–22
Christmas trees, 245–50
cigarettes, 130, 288
cities, 15, 16, 22, 50–51
composting in, 106, 108–9, 114, 121, 253
recycling in, 252–53
solid waste (MSW) from, 282–83
See also particular cities
Citywide Recycling Advisory Board (CRAB), 121, 252
class, 21, 22, 36, 40, 84, 245, 258
See also poverty
Clean Air Act, 24, 75
Clean Water Act, 42, 75, 100, 204, 219
Cloaca (installation), 231–32
Coca-Cola Company, 186, 189, 287
Cohen, Steven, 80
Colorado, 128, 213, 220
commercial waste, 3, 6, 64, 78, 88
Commoner, Barry, 166
composting, 17, 94, 105–26, 225, 278–79, 280
of Christmas trees, 246, 250
in cities, 106, 108–9, 114, 121, 125, 253
community-based, 116–17, 119
equipment for, 106–7, 109, 112, 124
in Europe, 125
of food waste, 15–16, 109, 114–16, 122, 125–26, 268–72
in landfills, 89–90
leachate from, 117–18, 119
of organic waste, 105, 108, 113, 259, 268–72
and oxygen, 109–10
politics of, 123, 126
of residential waste, 109, 111–12
of sewage, 229–30
of sludge, 109, 120–21, 133, 222
in toilets, 122, 227–29
vs. WTE, 80, 81
of yard waste, 107, 108, 109, 112, 116, 267, 268–72
and Zero Waste, 253, 255
computers, 165–75
Connecticut, 183, 292
Connor, Shayne, 222
consumerism, 18, 124, 193, 236, 239, 242, 294
consumption, 243, 255, 257, 282–83, 284
vs. production, 238, 239, 241–42
Container Recycling Institute, 155, 177, 183, 185
costs
of composting, 114, 119
of exporting waste, 72, 286
of labor, 260, 262
and landfills, 52, 58, 63, 64, 75, 76, 82
of private haulers, 70, 71–72
and privatization, 68
of recycling, 114, 173, 260, 262–63, 275
of transfer stations, 44
of WTE, 79, 81
Cowan, Rick, 71
Cradle to Cradle (McDonough and Braungart), 122, 239, 240, 256
credit cards, 235–36

cremation, 279
CSO (combined sewer overflow), 203–4, 222, 226, 287, 288

Dalton, Pam, 110–11
Darwin, Charles, 118
Datz-Romero, Christina, 116–19, 120, 121, 228, 270
Davis, Adam I., 283
decomposition, 89–92, 112, 114–15, 208, 231
DeCorti, Espera, 184
Delaware, 183
Dell, Michael, 173
Delvoye, Wim, 232
developing countries, 156, 162, 167, 193, 227, 229–30
  export of waste to, 132, 153, 155, 169–74, 187–89
Diamond, Jared, 285
diapers, disposable, 58, 197–98, 285
digesters, anaerobic food, 114–15, 119, 122–23, 245, 267, 271, 285
Diggins, Dennis, 85–95, 97–98, 100–101
dioxin, 189, 219–20, 221, 236
disease, 113, 207–8, 217–18, 222–23, 242
  See also asthma; carcinogens
disposable products, 21, 176, 197–98, 285, 290
Donato, Sam, 64–66, 67, 68, 73
Dutchin, Daren, 179–82

Earth Day, 3, 127, 146, 287
education, public, 8, 30, 253, 261, 278
electronic waste (e-waste), 165–75, 262, 267, 293
energy, 64, 115, 145, 161, 190, 242, 279
  electrical, 78, 135, 283
  and paper sludge, 133–34
  and recycling, 135, 139, 275, 293
  tidal, 254
  See also fuel
England, 51, 74, 92, 230
Environmental Defense Fund, 185, 235, 237, 260
environmentalists, 18, 24, 42, 43, 143–44, 148, 165
  in community, 52, 68, 76, 80, 81, 106
  and corporations, 184–85, 282, 287
  and credit cards, 235–36
  vs. industry, 252–53
Environmental Protection, New York Department of, 121, 164, 199, 200, 210, 219
Environmental Protection Agency (EPA), 79, 115, 124, 171, 198, 210, 292
  on composting, 111–12
  on dioxin, 219–20
  and fertilizer, 211, 222
  and gas collection systems, 92, 93
  and landfills, 51, 56–58, 59, 60
  National Priorities List of, 75
  and Part 503, 211, 216, 217
  on plastic, 190, 191
  and recycling, 127, 128
  Resource Conservation and Recovery Act of, 51, 75, 160–61
  and sludge, 221, 223
  Toxics Release Inventory of, 166
Erickson, Pete, 187
Eschenroeder, Alan, 81
European Union (EU), 125, 160, 173, 236–37
exports, 72, 246, 286
  of e-waste, 171, 172, 174
  of hazardous waste, 169–70
  of paper, 132, 187
  of plastic, 187–89, 264
  of scrap metal, 146, 153, 155
Exxon oil spill (1990), 53

fabric, 15, 17–18, 75, 97, 190, 273
Fat of the Land (Miller), 69, 290
Feliciano, Angel, 169, 170, 174
fertilizer, 109, 289–90
  classes of, 213, 215, 224
  and fossil fuels, 124–25
  labeling of, 222
  nitrogen in, 125, 221, 224
  sludge, 117, 211, 212–19, 221–26
  toxins in, 211, 216, 220
fires, 92–93, 197, 248
fish, 6, 8, 9, 56, 60, 210–11, 219
Flood, Mickey, 44, 47, 64, 66, 72, 73

Florida, 11, 128, 216, 223, 226
food waste, 36, 261
  anaerobic digesters of, 114–15, 119, 122–23, 245, 267, 271, 285
  composting of, 15–16, 109, 114–15, 116, 122, 125–26, 268–72
  disposers for, 109, 120–22, 123
  as percent of waste stream, 21, 124
Foote, Owen, 6
forests, 135, 136, 137, 139
Fort Tilden beach (Queens), 287–89, 290
Fossella, Vito, 100
France, 220, 254
Freilla, Omar, 218, 219
Fresh Kills Sanitary Landfill (Staten Island), 5, 42, 49, 50–62, 217, 245
  capacity of, 53, 239
  Christmas trees in, 246, 249, 250
  closing of, 11, 13, 44, 69, 96, 100
  composting in, 108, 114
  leachate from, 94–95
  opening of, 23–24
  postclosure, 85–101
  restoration of, 88–89
  tidal creeks at, 88, 92
  violations by, 99
  and World Trade Center, 100–101, 154
Freud, Sigmund, 231
fuel
  biodiesel, 201–2, 244, 257
  fossil, 4, 8, 20, 124–25, 275
  methane as, 93–94, 115, 119, 205, 207, 208
  and WTE, 77–81, 161, 286

Garbage Project (U. of Arizona), 13, 16, 51, 58, 89, 92, 109, 198
Garbology, National Institute of (NIG), 35
Georgia, 202, 222
Gere, Richard, 68
Germany, 125, 186, 262
Gibaldi, Lou, 205, 206–7
Gillespie, Spike, 239
Gitlitz, Jenny, 155
Giuliani, Rudolph, 13, 70–71, 72
glass, 15, 36, 91, 169, 180, 190
  from computers, 169, 172
  markets for, 278, 293
  recycling of, 5, 16, 33, 128, 129, 236, 257–58, 266, 273, 274, 278
  suspended recycling of, 14, 36, 59, 72, 154–55, 179–80, 251, 252, 275, 279
  See also bottles
Gleason, Phil, 62, 94–95
global warming, 44, 90, 121, 226, 275
  See also greenhouse gases
Gobé, Marc, 243
Gokaldis, Virali, 262–63, 268, 271
Goodstein, Judy, 129, 132, 134, 135
government, 142, 284
  federal, 24, 93–94
  local, 81
  state, 6, 24, 93–94
  subsidies from, 93–94, 139, 171, 255
  See also particular departments
Gowanus Canal, 3–10, 199–200, 225, 226, 287, 288
Grandner, Bill, 209
greenhouse gases, 74, 115, 125, 145, 177, 242, 270
  and composting, 119
  from landfills, 89, 93
  and paper, 136, 139
Greenpeace, 187, 188, 189, 236
green products, 237, 242–43
green roofs, 226, 285
Greenspan, Alan, 145
greenwash, corporate, 184–85
groundwater, 51, 57, 59, 60, 64, 75, 76

Haley, Robert, 253–63
Hanly, Michael, 163–64
Harrison, Ellen, 221
Hawaii, 143, 183
Hawken, Paul, 239, 283
Hawking, Stephen W., 70
hazardous waste, 43, 73, 75, 136, 158–75, 191, 275
  drop-off sites for, 163–65
  exports of, 169–70
  household, 158–59, 163, 278
  in landfills, 68–69, 76, 159, 160, 161
  in leachate, 58–60
  and sludge, 211, 223
Heckler, Phil, 198–209

Hershkowitz, Allen, 137–39, 253
holidays, 235, 238–39, 244, 245–50
horses, 22, 289–90, 292
Hudson, Don, 201–2
The Humanure Handbook (Jenkins), 229

Iceland, 156
Illinois, 31, 52, 212
incineration, 16, 40, 51, 145, 185, 249, 286
  and employment, 284
  of e-waste, 166, 174
  government support for, 23, 93–94, 171
  and hazardous waste, 160, 161, 164
  opposition to, 24, 187
  out-of-state, 44, 53
  and paper, 133–34, 141
  of plastic, 191
  of sludge, 212
  in WTE plants, 77–81
India, 162, 169, 187–89
Indiana, 135, 162, 203
industry, 64, 184–85, 282, 287
  and bottle bills, 183–85
  computer, 172, 173
  vs. environmentalists, 252–53
  paper, 133–35
  plastics, 181–82, 186, 190–91, 193, 194
  and recycled content, 186, 242, 244, 262, 266, 274, 276
  siting of, 6, 212–13
  sludge, 220, 223
  sustainable, 238
  waste from, 3, 6, 64, 219, 239, 253, 275, 283
  and Zero Waste, 255, 257
  See also mining
insects, 36, 107, 110, 269, 278–79
  See also vector control
Institute for Local Self-Reliance, 80–81, 284
Integrated Environmental Services Incorporated (IESI), 39–40, 44–49, 53, 71
  Bethlehem landfill of, 63–68, 75–76, 81, 84, 158
International Coastal Cleanup, 287–89, 294
Internet, 112, 171, 286
Iowa, 128, 185, 264
Ireland, 193, 254
Istrefi, Adrian, 246–49
Italy, 236–37

Jamaica Bay (New York), 22, 56, 289–94
Jenkins, Joseph C., 229
Jepson Prairie Organics, 268–72
Jersey City (New Jersey), 146
Johnson, Austin, 140

Keep America Beautiful (KAB) campaign, 184–85
King, Carol, 222
Klima, Ivan, 39
Kloor, Keith, 44, 98
Knapp, Dan, 192, 273–74
Krupnik, Tim, 227–29, 230
Kwiatkowski, Ron, 178–79, 181

labor, 10, 238, 260, 262, 263–64
labor unions, 253
landfills, 6, 50–62, 121, 185, 243, 284
  active face of, 63–84
  bioreactor, 57–58
  capacity of, 286
  composting at, 89–90, 108, 114, 122, 259
  contents of, 145, 153, 155–56, 180, 182, 190, 191, 197–98, 212, 239, 246
  costs of, 52, 58, 63, 64, 75, 76, 82
  covers for, 23, 59, 97
  decomposition in, 89–92, 231
  diversion rates from, 23, 253, 259, 262, 264–67
  dry-tomb, 56–57
  and environment, 51, 74, 76
  and EPA, 51, 56–58, 59, 60
  e-waste in, 165, 166, 172
  government support for, 93–94, 171
  greenhouse gases from, 89, 93
  and groundwater, 51, 57, 59, 60, 64, 75, 76
  hazardous waste in, 68–69, 75, 76, 159, 160, 161, 162, 220
  host fees for, 40–43, 76
  impacts of, 56–58
  layers in, 66, 74–75, 97
  leachate from, 51, 57–58, 65, 75, 94–95
  liners for, 57, 59, 75
  marine, 48
  mega-, 42, 52, 238
  mining of, 156
  on Nantucket, 122–23
  in New Jersey, 81–84
  out-of-state, 44, 47, 52–53, 77

landfills (*cont.*)
  and packaging, 236
  and paper, 128, 133, 135–36, 138–39, 141
  privatization of, 68
  profits from, 52, 265
  vs. recycling, 65, 189, 263
  restoration of, 88–89
  sanitary, 23, 51–52, 89
  secrecy about, 72–74
  and sewage, 138–39, 198, 205, 206, 216
  siting of, 40, 43
  stabilization of, 91, 92
  and WTE plants, 79
  vs. Zero Waste, 256
  See also Fresh Kills Sanitary Landfill
Lange, Robert, 72, 246, 250
laws. *See* regulations
leachate
  and composting, 117–18, 119
  from landfills, 51, 57–58, 65, 75, 94–95
  nutrients in, 60
  toxins in, 58–60, 98, 99, 162, 191
  treatment of, 98–99
Lee, G. Fred, 57–58, 60, 92, 93
Leonard, Ann, 187–89, 190
Leopold, Aldo, 278
Lethem, Jonathan, 5
Lewis, David, 223
Light, Andrew, 288–89
Lin, Maya, 137
Long Island (New York), 37, 44, 128
  See also American Ecoboard
Los Angeles (California), 31, 109, 128, 143, 212
Love and Garbage (Klima), 39
Lovett, Chris, 132–35
Lower East Side Ecology Center (New York), 115–19
Lydgate, Chris, 35
Lykes Brothers Citrus, 223–24

MacBride, Samantha, 281–83
MacDowell, Andie, 128
Magdits, Lou, 169
Mailer, Norman, 35
Maine, 128, 173, 183, 185, 201–2, 264
Manhattan, 33, 132, 203
Manning, Richard, 125
Marcalus, Peter, 138
markets, 122, 162, 242, 243, 265–66, 286
  for e-waste, 171, 175
  for glass, 278, 293
  for paper, 132, 137–38, 140
  for plastic, 187, 264
  for scrap metal, 145, 153
Martinez, Eve, 171
Maryland, 212
Massachusetts, 77, 122–23, 127, 172, 202, 211, 245, 285
materials recovery facilities (MRFs), 71–72, 128, 183, 253, 260, 274, 286
  and plastic, 179–82
  in San Francisco, 254, 263–66
Matsil, Marc, 96
May, William F., 185
McCarthy, Fishhooks, 22
McDonald's, 237
McDonough, William, 122, 239, 240, 256, 283
McLean, Jack, 129–31, 133
Meadowlands (New Jersey), 56, 58, 60, 77, 92
meat, 19, 110–11, 244, 271
medical waste, 42, 58, 207–8, 220, 271
mercury, 58, 136, 152, 153, 158, 161–64, 211, 219, 279
metals, 15, 18, 47, 190, 283
  from computers, 168, 169
  copper, 58, 146, 152, 160, 172, 175, 219
  exports of, 187
  household, 58, 142
  iron, 63, 146, 153
  in leachate, 58, 98, 99
  lead, 44, 136, 153, 166, 169, 219, 221, 256
  nickel, 159, 160
  as pollutants, 58, 98, 145, 159, 166, 203, 219, 220
  precious, 160
  recycling of, 14, 33, 128, 129, 142–57, 180, 181, 236, 251, 258, 259, 273, 278
  steel, 64, 143, 145, 146, 153, 154, 181
  virgin, 160
  and WTE plants, 77–78, 79
  See also aluminum; mining
methane, 60, 74, 89, 90, 121, 139
  collection of, 51, 57, 92–94
  fires from, 92–93
  at Fresh Kills, 88, 89, 92

as fuel, 93–94, 115, 119, 205, 207, 208
methylmercury, 136, 162
Michigan, 11, 41, 203
Miller, Benjamin, 69, 290
mining
of e-waste, 175
of landfills, 156
pollution from, 160, 166–67
subsidies for, 139, 171
of virgin metals, 160, 163
waste from, 275, 283
Minnesota, 126, 128, 259, 264
Mississippi, 294
Mitchell, Joseph, 56
Mobro (garbage barge), 127–28, 239
mongo, 37, 38, 155
Montague, Peter, 255
Moore, Charles, 192
Morgante, Frank, 45–46
Morgenthau, Robert, 69–70, 71
Moses, Robert, 23–24, 290
MRFs. *See* materials recovery facilities
municipal solid waste (MSW), 282–83
Murphy, Billy, 27–40, 130–31, 244, 252, 278
Murphy, Cullen, 69
Murray, Chris, 82–83

Nagle, Robin, 69, 85–87, 95–97, 100, 277, 282
Native Americans, 5, 184
Natural Capitalism (Hawken), 239
Natural Resources Defense Council (NRDC), 136, 183, 203, 253, 260, 262, 268
Netherlands, 125, 186
Neu, Hugo, 142–43, 144, 149, 181, 264
Neu, Wendy, 143–44, 154, 156, 194
Neu Corporation, 142–57, 278, 293
New Jersey, 44, 53, 54, 162, 164
and interstate garbage trade, 52, 63
metal recycling in, 143, 146
Monmouth County in, 81–84, 87–88
waste from, 41, 42, 78
and WTE plants, 77–80
See also Arthur Kill; Meadowlands
New Orleans (Louisiana), 50
New York City
art in, 231–32
composting in, 108, 115–19, 121
diversion rates in, 253, 259, 262
ecological footprint of, 10–11
green roofs in, 226
hazardous waste in, 159, 172
landfill for, 42–43
Parks Department of, 114
pigs in, 112–13
recycling in, 128, 129, 261, 265
rivers of, 6, 8, 116, 118, 203
vs. San Francisco, 253, 259, 262
sanitation districts in, 30–31
sewage treatment plants in, 199, 204–9
sludge fertilizer from, 117, 212, 215
transfer stations of, 43–49, 139–40
trash collection in, 24, 260, 281, 285, 286
waste stream from, 11, 20, 21–22, 41, 42, 76–81
See also Bronx; Brooklyn; Manhattan; Queens; Sanitation, New York Department of; Staten Island
New York Organic Fertilizer Company (NYOFCo), 212–19, 221–25, 278
New York State, 11, 52, 64, 69, 252, 292
bottle bill in, 182–83, 185
sewage in, 198–209, 211
See also Environmental Protection, New York Department of; Jamaica Bay; Long Island
New Zealand, 254
Nichols, Mike, 176
Nike (firm), 256
nitrogen, 99, 121, 228, 271, 275
in fertilizer, 125, 221, 224
in humanure, 229, 231
from sewage treatment plants, 291–92
nonhazardous industrial waste (NIW), 282–83
Norcal, 254, 256, 259, 262, 264, 265, 266, 268, 274

North Carolina, 58, 153, 255
Norton, Gale, 51–52
Norway, 11
NYOFCo. *See* New York Organic Fertilizer Company

oceans, 23, 125, 191–92
  sewage in, 210–11, 291–92
odor, 110–11, 116, 218
Ohio, 52, 53, 162, 203, 216, 226
oils, 194, 201–2, 289
  See also petroleum products
Oklahoma, 214
Onuska, John, 160
Oregon, 35, 125, 128, 183, 264, 266
organic products, 222, 242, 271–72
organic waste, 58, 115, 125, 258
  composting of, 105, 108, 112, 113, 259, 268–72
  as percent of waste stream, 124
  putrescible, 22, 33, 112, 192, 273
  recycling of, 273, 278
  See also food waste; yard waste
organized crime, 69–72
OSHA (Occupational Safety and Health Administration), 143, 191
Outerbridge, Tom, 113–15, 119, 120–21, 249, 271
Owls Head sewage treatment plant (Brooklyn), 200–209, 210, 225, 231

Pabon, Gloria, 248
packaging, 16, 184, 190, 236–37, 261, 274
  food, 280–81
  paper *vs.* plastic, 193, 244
  reusable, 262, 284
  in waste stream, 21, 124, 275–76
paper, 15, 235
  composting of, 112, 114, 115
  exports of, 132, 187
  manufacture of, 18, 133–35
  in packaging, 193, 236–37
  as percent of waste stream, 20, 109
  vs. plastic, 193, 244
  post- *vs.* preconsumer, 134, 137–38
  sludge from, 133–34
  toilet, 138–39
  virgin, 133, 136–37, 139
paper recycling, 14, 33, 127–41, 236, 237, 267, 273, 278
  and contamination, 274
  and DSNY, 132, 139–41
  and landfills, 128, 135–36, 138–39, 141
  markets for, 132, 137–38, 140
  in pay-as-you-throw system, 258
  and san men, 129–31
Parkinson, C. Northcote, 16
"Pavlov's Pack Rats" (Bleifuss), 275
pay-as-you-throw system, 258–63
PCBs, 76, 153, 219, 221
Pennsylvania, 47, 52, 202
  landfills in, 41, 63–68, 75–76, 81, 84, 99, 100, 158
  See also Bethlehem
Pepsi-Cola Company, 186, 187, 189
Perkins, Tony, 35
Perrani, Mike, 278
Per Scholas (computer recycling firm), 168–69, 170
petroleum products, 53, 135, 145, 164, 171, 190, 203, 283
  See also fuel
pigs, 50, 112–13, 120, 282
plastic, 16, 36, 156, 176–94, 258, 288
  biodegradation of, 91, 191, 286
  from computers, 168, 169, 172
  EPA on, 190, 191
  exports of, 187–89, 264
  externalities of, 192–93
  introduction of, 21, 45
  markets for, 187, 264, 269
  in packaging, 236
  vs. paper, 193, 244
  particular types of, 177, 181–82, 186–89, 192, 236, 241, 253, 264
  pollution from, 187–92
  toxins in, 59, 99, 190–91
  virgin, 182, 185, 186, 188, 189
plastic bags, 192–93, 280, 281, 286
plastic recycling, 33, 128, 236, 255, 273, 278
  in MRFs, 179–82, 264
  rates of, 177

suspension of, 14, 36, 59, 72, 154–55, 179–80, 251, 252, 275, 279
population growth, 11, 124, 242
poverty, 15, 21, 43–44, 212–13, 245, 258
privacy rights, 34–35
private haulers, 69–70, 71–72, 246, 247–48
producer responsibility, 253, 255, 262, 274, 276, 284, 294
extended (EPR), 173
Prolerizer, 149–51, 152, 155, 157, 181
Pryor, Greg, 75, 268–72

Queens (New York), 143, 201, 287–89, 290

racism, environmental, 43, 80, 213–15
See also African Americans
Racketeering in Legitimate Industries (Reuter), 69
radioactivity, 158, 220–21
rainwater. *See* storm runoff
Rampton, Sheldon, 211
Rathje, William, 13, 51, 69, 91, 94–95, 240, 277
rats, 36, 51, 116
See also vector control
recycled content, 186, 237, 242, 244, 262, 266, 274, 276
recycling, 105, 122, 241
categories for, 273–74
compliance with, 130, 140–41
and contamination, 274
cost effectiveness of, 114, 173, 260, 262–63, 275
curbside, 127, 140–41, 189, 236, 252, 260, 280
decline of, 82, 114
dual *vs.* single stream, 128, 129, 253, 266
education on, 30, 239, 253, 261
employment in, 137, 182, 188–89, 275, 284
and energy, 135, 139, 275, 293
of e-waste, 171–75
and government, 94, 171
individual, 281–82
vs. landfills, 65, 189, 263
markets for, 265–66
municipal, 252–53
suspension of, 14, 36, 59, 72, 154–55, 179–80, 251, 252, 275, 279
of organic waste, 123, 200–201, 273, 278
rates of, 128, 136, 177, 236, 264–65
regulations on, 30, 262, 266, 274, 276
symbol for, 177, 181–82
total, 273
and trade, 18, 52, 72
value of, 243–44, 281–83
waste-to-energy, 80, 81, 174
and Zero Waste, 253–75
See also aluminum; glass; metals; paper recycling; plastic recycling
"Recycling Is Garbage" (Tierney), 260
Rees, William, 10
regulations, 81, 99, 124, 128, 173, 283
and landfills, 52
recycled-content, 266, 274, 276
on recycling, 30, 262, 266, 274, 276
violations of, 30, 34, 63, 70, 71, 86, 99
Reichel, Andy, 81–82
residential waste, 11–16, 142, 219, 279–80, 282
vs. commercial waste, 78, 283
composting of, 109, 111–12
hazardous, 58, 158–59, 159, 163, 164, 165, 278
Reuter, Peter, 69
"Round River" (Leopold), 278
rubber, 18, 256, 265, 273, 274
Rubbish! (Rathje and Murphy), 69, 94

salt marshes, 53, 55, 56, 60–62, 95
San Francisco (California), 125, 128, 156, 253–75
MRF in, 263–66
vs. New York City, 253, 259, 262
pay-as-you-throw system in, 258–63
Sanitation, New York Department of (DSNY), 27–35, 44, 72, 156, 245, 246

Sanitation (*cont.*)
and composting, 106, 114
and Fresh Kills, 85–92
and paper recycling, 132, 139–41
sanitation districts, 30–31
sanitation workers (san men), 14, 27–49, 73, 159, 249, 254, 281, 285
female, 248
gestures of, 155
hazards for, 32–33
and odor, 111
and paper recycling, 129–31, 140
and privacy, 34–35
studies of, 39, 85–86
Sanjuro, Eva, 219
Scandinavia, 186
scavenging, 15, 19–20, 21, 45, 50, 73, 75
for deposit bottles, 36, 251–52, 259, 278
in India, 187
for mongo, 33, 37, 38, 155
and privatization, 68
Scorziello, Peter, 213–21, 224–25
Scotland, 254
Seattle (Washington), 108, 114, 125, 187, 266–68, 294
September 11, 2001 attacks. *See* World Trade Center
Serrano, José, 222
Seventh Generation (brand), 138, 243
sewage, 3, 6–8, 60, 197–209
history of treatment of, 202–3
pharmaceuticals in, 207–8
pollution from, 210–11
and storm runoff, 7, 203–4, 222, 226, 287, 288
See also sludge
sewage treatment plants, 73, 138, 198–209
in New York City, 199, 204–9
nitrogen from, 291–92
sludge from, 100, 206–9, 210–32
Sheehan, James, 154–55
Shinn, Steve, 146–57
shoes, 256, 293
Sierra Club, 3, 185, 253
Silicon Valley Toxics Coalition (SVTC), 166, 169–70
Simon, Neil, 35
sludge
composting of, 109, 120–21, 133, 222
in fertilizer, 211, 212, 216, 222, 223, 226
on food crops, 213, 222, 225
and food waste disposers, 120, 121
from Fresh Kills, 99–100
from paper, 133–34
recycling of, 273
from sewage treatment plants, 206–9, 210–32
toxins in, 211, 219, 220, 223, 256
and vector control, 217–18
and wetlands, 226–27
Smutko, Harry, 209
source reduction, 239–40, 242, 286
South Africa, 193
South Carolina, 41, 53
Springsteen, Bruce, 131
Spurlock, Morgan, 237
Staten Island (New York), 23–24, 100–101, 129–35, 132
See also Fresh Kills Sanitary Landfill
"The State of Garbage in America" (*BioCycle* magazine), 93, 108, 128
Stauber, John, 211
Steinberg, Ted, 112–13
Steisel, Norman, 285
Stenshol, Shaun, 257–58
storm runoff, 7, 203–4, 222, 226, 287, 288
Strasser, Susan, 16, 21, 124, 176
Sullivan, John, 27–40, 130–31, 247, 252, 278
Superfund sites, 52, 75, 166, 220
Super Size Me (Spurlock), 237
Supreme Court, US, 35, 73
sustainability, 238, 243, 281–82, 284
Sustainable South Bronx, 218–19
Switzerland, 173
Synagro, 89–90, 213, 222

Takedown (Cowan), 71
Tammany Hall, 22, 23
taxes, 93–94, 139
technology, 21, 238, 284–86, 293
Tennessee, 162
Terlizzi, Jerry, 28, 29–30
Texas, 11, 212, 214–15, 239
Themelis, Nick, 80–81
Thomas, Lenny, 5
Tierney, John, 260, 275

toilet paper, 138–39
toilets, composting, 122, 227–29
Toro, Jorge, 129–31, 133
TOWEL effect, 135
Toxic Sludge Is Good for You! (Stauber and Rampton), 211
toxins, 190, 242, 284, 288
  in fertilizer, 211, 216, 220
  in leachate, 58–60, 98, 99, 162, 191
  in sludge, 211, 219, 220, 223, 256
  See also hazardous waste
trade, 18, 19–20
  interstate, 52–53, 63
  See also markets
Trade Waste Commission, 70–71
transfer stations, 39–40, 43–49, 69, 71, 139–40
Turso, Vito, 285
Tuscano, Don, 98–99

Ukeles, Mierle Laderman, 99
United Nations (UN), 229, 238
Urban Ore (Berkeley, California), 192, 272–74

Vanderbilt, Gloria, 35
vector control, 124, 217–18, 225, 269
Vermont, 183, 184, 185, 226
Vincenz, Jean, 51
Virginia, 41, 52, 53, 58
Visy Paper, 129–35, 140

Wackernagel, Mathis, 10
Wade Salvage, 164–65
Walker, Nigel, 271–72
Walsh, Daniel C., 20–21
Waring, George E., Jr., 22–23, 125
Warren, Rebecca, 267–68
Waste and Want (Strasser), 16, 21, 124, 176
Waste Management (firm), 40, 41, 67, 71–72, 75, 256, 268, 283
waste stream
  components of, 20, 21, 109, 124, 275–76
  per capita, 11, 20, 279
  seasonal variations in, 36–37, 244–45
waste-to-energy (WTE) plants, 77–81, 286
wastewater, 207–8, 220
  treatment plants for, 6, 7, 42, 43, 58, 109, 120–22, 219, 226, 272
water, 6, 121, 122, 125, 145
  bottled, 177
  clean, 18, 42, 75, 100, 204, 219
  and paper, 135, 136, 137, 139
  quality of, 8, 9, 199
  and sewage, 229–30
waterways, 6, 8, 53, 54, 118, 197
  sewage in, 3, 42, 199–200, 202–3
  See also Arthur Kill; Gowanus Canal; Jamaica Bay; oceans
Weberman, A. J., 35
Weissman, Arthur, 242
West Virginia, 42–43
wetlands, 6, 23, 95, 97–98, 227–28
White, Christopher, 41
White, Jim, 77–79, 161
wildlife, 167, 191–92, 207–8, 214–15, 231, 240–41
Wilson, Edward O., 238
Wintour, Anna, 117
Wisconsin, 127, 212, 222
World Trade Center, attacks on, 100–101, 153–54
Worldwatch Institute, 137, 174
worms, 108, 112, 115, 117–19, 228, 267
Wurth, Al, 42, 68

yard waste, 123, 124, 258
  composting of, 107, 108, 109, 112, 116, 267, 268–72

Zero Waste concept, 148, 251–76, 284
Zimmer, Happy, 56

About the Author

Elizabeth Royte is the author of *The Tapir's Morning Bath: Solving the Mysteries of the Tropical Rain Forest,* and has written for the *New York Times Magazine, Harper's, National Geographic, Outside, Smithsonian, The New Yorker,* and numerous other national magazines. A former Alicia Patterson Foundation Fellow, she lives in Brooklyn with her husband and their daughter.

GUIDE
READERS' PICK
INSIDE

Reading Group Guide

Garbage Land

ON THE SECRET TRAIL OF TRASH

Elizabeth Royte

A conversation with Elizabeth Royte

What prompted you to explore the hidden world of garbage?

I had long been curious to know where my garbage went, and what became of it. I knew that ever since New York City's local landfill, on Staten Island, had closed, my stuff was going farther and wider than many other people's garbage. I was also dissatisfied with the answers I got when I asked sanitation workers where my plastic bottles and cans were going. "They get recycled," they'd tell me, usually with a shrug, but they didn't know where, or into what.

What are your thoughts on home recycling?

Keep doing it! Keep setting aside whatever your community will take. And find outlets for other things that aren't accepted on the curb, like batteries and cell phones, but are toxic in landfills and incinerators. Then go out and buy products made from postconsumer recycled content: you'll be closing the loop, avoiding some pollution and energy use upstream, and signaling the market that this stuff has value.

Now, having said that, I'll also say this: don't drive yourself crazy with every potentially recyclable thing in your trash can. There are more important things you can do for the environment than fuss over your rubber bands or lightbulbs. For example: think about what you drive (its mileage), about how you heat or cool your home, and whether you could eat less meat.

What was the most difficult (or disgusting) thing you did in researching this book?

I have a high gross-out threshold, so even handling biosolids (dewatered sewage sludge) in a manufacturing plant wasn't as difficult as, say, trying to pry information from people in the waste world who had no interest in giving me information, access, or their opinions.

Did your research ever leave you smelling outrageously bad?

Funny you should ask. I toured the biosolids plant over several hours on a cold winter day, when I was wearing a fleece vest and a down jacket. You couldn't find better materials for absorbing odors. When I sat down on the subway to go home and opened my notebook, I found that with every movement of my arms I released a puff of eyewateringly bad air. People were actually moving away from me, giving me dirty looks. I had to ride three different trains back to Brooklyn: it was the longest commute of my life.

What did you learn that shocked you the most?

I learned that for every barrel of trash you set on the curb, there are 71 barrels of waste generated upstream, in manufacturing all this stuff you bought, used for just a short amount of time, and then consigned to the dump. Municipal solid waste is only 2 percent of the entire U.S. waste stream: the rest is manufacturing waste, agricultural waste, mining waste, construction and demolition debris, and other rarefied categories. But that 98 percent isn't unrelated to the two: it's all the waste that goes into the products and processes that keep us alive and happy. If we reused and recycled more, the numbers associated with extracting virgin materials (like wood, metals, and oil and gas) and transforming them into consumer goods would go down.

You came across quite a few interesting waste reduction solutions in your journey (e.g., humanure and desktop composters). Would you ever consider using any of these?

I generate far too many kitchen scraps (a little over twenty-two pounds a month) and have far too little counter space to make use of a tabletop worm bin. I'm not really interested in making humanure, with worms and newspaper (the way Tim Krupnik, in my book, did). But I'd be delighted to have a composting toilet. They don't use any water, they don't smell, and they turn human waste into fuel for soil.

What do you think about the steps taken by cities such as San Francisco and Seattle to reduce waste?

I think those cities are bold and forward thinking. They're experimenting with collection systems, they're requiring manufacturers to use recycled content, they're developing markets for recycled goods, and they're challenging business as usual (that is, landfilling and incineration). I especially like the household and commercial collection of food scraps. According to the EPA, 60 percent of landfill contents is potentially compostable. That's a lot of space that could be freed up, and it's a lot of organic material that wouldn't generate greenhouse gases or leachate because it's not buried.

Did researching and writing this book change your outlook on garbage?

Yes. I came to see that municipal collection of garbage, which started as an issue of health and commerce (merchants couldn't move consumer goods if the streets were impassable), is just another way for producers and manufacturers to externalize the social and environmental costs of their product waste. So long as taxpayers support the burying and burning of product waste (as opposed to organic waste, which can be composted and returned to the earth), producers will continue to pump out goods that fall apart, rapidly become obsolete, can't be repaired or recycled, that are toxic, and that will be buried or burned at taxpayer expense. If manufacturers were required to take responsibility for their products (including packaging and containers), they'd have an incentive to make things that can be reused or recycled or composted, and that contain fewer hazardous materials.

How did your family respond to your decision to examine garbage?

Admirably. My daughter thought pawing through the trash was fun (for a while at least, then she got bored; also, converting ounces to decimal fractions left her cold). My husband was as supportive as possible. I believe that when he periodically dumped coffee grounds into the trash, instead of into the compost bucket, it was an oversight. He helped me by using string bags instead of plastic, by forgoing the use of paper towels, and by refusing to buy things, such as plastic toys with moving parts, that would mess up my weekly average (I was trying to throw away less, by weight, than the average American).

What do you want people to take away from reading your book?

Just a few things. I want people to be aware that garbage doesn't go away, that it isn't environmentally benign, that there are things individuals can do to lessen their environmental impact (go to www.garbageland.us and click on "What to Do" for some ideas), and that recycling is good, but fundamental changes in how manufacturers design and make consumer products is even more important.

Questions and topics for discussion

1. Elizabeth Royte's trip to the Gowanus Canal was one of the factors that compelled her to embark on her strange journey. What compelled you to pick up *Garbage Land* and join her on her journey? Did you ever wonder what actually happened to the things you throw away?

2. *Garbage Land* is a first-person narrative in which Elizabeth Royte guides readers through the hidden world of garbage. What does this first-person point of view add to the book? Do you think this book would have been as effective if it had been written from an omniscient narrator's point of view?

3. The author notes that trying to be environmentally responsible isn't always easy. Indeed, she writes (on page 141) that her publisher couldn't justify the added cost of printing the book on recycled paper stock. How can we reconcile the moral arguments for recycling with the economic arguments against it?

4. *Garbage Land* features a bevy of interesting characters, many of whom suggest innovative ways to reduce our garbage footprint. For example, Christina Datz-Romero is a proponent of desktop worm composters. Would you consider using one of these innovative methods? If so, which ones?

5. Elizabeth Royte writes about trailing sanitation workers, visiting waste transfer stations, and exploring landfills. How did getting an inside look at these three parts of the disposal process broaden your understanding of the costs of consumption? What was most surprising about the author's visits?

6. Less than 27 percent of garbage is recycled and composted. But in some cities—San Francisco, for example—the rate is much higher. What do you think about the measures that the San Francisco government has taken to inch toward the "dream of zero waste"? Do you think similar measures could be effective in your city or town?

7. The author discusses recycling and reducing consumption. Which approach to minimizing garbage does she ultimately think is more effective? Do you agree?

8. Have you noticed a change in our country's waste stream in your lifetime? Explain your answer.

9. In *Garbage Land,* a former sanitation commissioner controversially claims, "In the end, the garbage will win." Does the author seem to agree with this sentiment? Do you?

10. How can we, as a society, reduce our garbage footprint? Has this book made you decide to change your behavior?

Elizabeth Royte's suggestions for further reading

Cradle to Cradle: Remaking the Way We Make Things by William McDonough and Michael Braungart
This book seems a little pie-in-the-sky in places, but mostly it's inspirational. Its central message is that it doesn't have to be this way: manufacturers *can* produce consumer goods without harming the earth.

Fat of the Land: Garbage in New York—The Last 200 Years by Benjamin Miller
Exhaustive, wry, eye-opening, especially if you've ever lived in New York City, waste capital of the world. Puts today's garbage follies into brutal economic and political context.

Rubbish!: The Archaeology of Garbage by William L. Rathje and Cullen Murphy
The urtext for anyone interested in what our trash reveals about us. Highly readable, and often funny.

Underworld by Don DeLillo
A fascinating and broad-ranging novel about a waste trader who imagines garbage as an evolutionary force, with the power to shape early man. You won't learn much about household waste here, but the landfill scenes manage to be both horrifying and uplifting.

Waste and Want: A Social History of Trash by Susan Strasser
Extremely useful for explaining the barter system of town and city, Strasser's history shines light on our path from producers to consumers.

Toxic Sludge Is Good for You! by John Stauber and Sheldon Rampton
An exposé of the public relations industry; particularly good on the regulatory makeover that turned contaminated sewage sludge into farm fertilizer.

Gone Tomorrow: The Hidden Life of Garbage by Heather Rogers
An unflinching look at the American way of consumption, a critique of recycling, and a brave call for upstream reform from government and manufacturers.

Takedown: The Fall of the Last Mafia Empire by Rick Cowan
Written by the undercover cop who helped Manhattan's district attorney run the mob out of New York's private carting business; reads like good crime fiction.

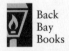